T0134945

Advanced Structured Materials

Volume 93

Common engineering materials reach in many applications their limits and new developments are required to fulfil increasing demands on engineering materials. The performance of materials can be increased by combining different materials to achieve better properties than a single constituent or by shaping the material or constituents in a specific structure. The interaction between material and structure may arise on different length scales, such as micro-, meso- or macroscale, and offers possible applications in quite diverse fields.

This book series addresses the fundamental relationship between materials and their structure on the overall properties (e.g. mechanical, thermal, chemical or magnetic etc.) and applications.

The topics of *Advanced Structured Materials* include but are not limited to

- classical fibre-reinforced composites (e.g. class, carbon or Aramid reinforced plastics)
- metal matrix composites (MMCs)
- micro porous composites
- micro channel materials
- multilayered materials
- cellular materials (e.g. metallic or polymer foams, sponges, hollow sphere structures)
- porous materials
- truss structures
- nanocomposite materials
- biomaterials
- nano porous metals
- concrete
- coated materials
- smart materials

Advanced Structures Material is indexed in Google Scholar and Scopus.

More information about this series at http://www.springer.com/series/8611

João M. P. Q. Delgado
Antonio Gilson Barbosa de Lima
Editors

Transport Phenomena in Multiphase Systems

 Springer

Editors
João M. P. Q. Delgado
Faculty of Engineering
University of Porto
Porto
Portugal

Antonio Gilson Barbosa de Lima
Department of Mechanical Engineering
Federal University of Campina Grande
Campina Grande
Brazil

ISSN 1869-8433   ISSN 1869-8441   (electronic)
Advanced Structured Materials
ISBN 978-3-030-08167-6   ISBN 978-3-319-91062-8   (eBook)
https://doi.org/10.1007/978-3-319-91062-8

Printed on acid-free paper

This Springer imprint is published by the registered company Springer International Publishing AG part of Springer Nature
The registered company address is: Gewerbestrasse 11, 6330 Cham, Switzerland

Contents

Hygrothermal Performance Evaluation of Gypsum Plaster Houses in Brazil

João M. P. Q. Delgado and P. Paula

Abstract Since the beginning of human civilization different strategies were used in order to build healthy, comfortable and economical houses. In recent years, the hygrothermal conditions of building's interior has been widely studied and parameterized. The use of gypsum plaster in building blocks, in Brazil, is still a restricted item, due to lack of knowledge on technology. Another factor that restricts the use of the blocks is the location of the deposits of gypsum, a material that gives rise to the plaster, which are located in the West of the State of Pernambuco, located far from the large consuming centers, potentially located in the southeast of the country. However, the interest in your study is determined by the annual consumption growth that reaches about 8% per year in Brazil. For the development of this study, in particular for Brazil, it is important to analysis the housing deficit in Brazil. The analysis indicates a clear need for new housing construction in the country. Preferably should be chosen simple constructive solutions, fast and easy to implement, without neglecting the comfort and durability. The use of local natural endogenous resources such as gypsum plaster, can be an attractive solution from an environmental point of view as well as economical. Brazil is a very large country with very different climates throughout its regions, so it was made a comprehensive study of the country's climate data and the constructive strategies adopted by NBR15220 and NBR 15575 standards, for each bioclimatic zone, to ensure the minimum comfort conditions. The hygrothermal analysis, reflected on the thermal comfort and night time ventilation in Brazilian gypsum plaster houses, was based on the adaptive model described in ASHRAE 55 and ISO 7730 standards for the evaluation of thermal comfort.

Keywords Gypsum plaster · Brazilian gypsum houses · Hygrothermal performance · Numerical simulation · Thermal comfort

J. M. P. Q. Delgado (✉) · P. Paula
Civil Engineering Department, Faculty of Engineering, University of Porto,
CONSTRUCT-LFC, Porto, Portugal
e-mail: jdelgado@fe.up.pt

P. Paula
e-mail: dec11021@fe.up.p

© Springer International Publishing AG 2018
J. M. P. Q. Delgado and A. G. Barbosa de Lima (eds.), *Transport Phenomena in Multiphase Systems*, Advanced Structured Materials 93,
https://doi.org/10.1007/978-3-319-91062-8_1

1

1 Introduction

Brazil is a country of continental dimensions, with 14,500,000 km2, and a population that already exceeds the 190 million inhabitants, according to statistical data from the last census conducted in 2010, by IBGE [1]. The construction of gypsum houses emerged, in Brazil, for over than one decade, and has aroused the interest of several researchers linked to the various branches of civil engineering.

The country has large reserves of gypsum plaster, more than 1.2 billion tons [2], with wide potential for exploitation and excellent level of purity, and reaching 95%. The use of this material for building construction, as well as their contribution to the constructive industry, appears as more an alternative material with full capacity of partial or total replacement of the materials currently used on large scale in construction industry. Gypsum blocks are a building material that promises to bring down the cost of construction by at least 30% in comparison with other materials such as drywall system, brick walls or concrete/cement block walls.

Thus, becomes important to recommend a study of this type of construction, as a way to increase this constructive mode in Brazil and in other countries, without, however, forgetting the minimum requirements of comfort and energy efficiency. The contribution of the study will bring positive consequences for regional and local economic growth. The study of thermal comfort and energy efficiency in this type of housing was one of the central objectives of this work.

João Pinheiro Foundation [3] estimates that the current Brazilian housing deficit is around 7 million units, with a methodology based on two distinct aspects: the amount of the housing deficit, scaling the stock of dwellings and the inadequacy of houses considering the internal specificities of this stock.

According to CEDEPLAR/UFMG [4], between 2007 and 2023, the demands for housing in this period, added to the already accumulated deficit of 7 million, culminated in a total of approximately 35 million of housing deficit. In the five largest metropolitan areas, with more than 1 million inhabitants, the population living in slums is very high: São Paulo—596,000, Rio de Janeiro—520,000, Belem —291,000, Salvador—290,000 and Recife—249,000 people.

Based on this statistical data, the motivation for a study that presents a gypsum house as an option to mitigate the issue of Brazil's housing deficit is very attractive, assuming that the buildings built in plaster may present potential habitability levels, thermal comfort and low energy consumption. However, it should be noted that other behaviours should be examined and analysed in detail, in future works, such as: water influence and pathologies associated, durability, acoustic behaviour, maintenance, etc.

This work is divided in several sections presented as the following:

- Chapter "Influence of Reinforced Mortar Coatings on the Compressive Strength of Masonry Prisms" presents a state of the art related to the problem of housing deficit in Brazil, a projection of housing demands, the Brazilian housing policy after 2003, discusses the policy of house promotion by the Brazilian Government in the last years, presents an exhibition of energy policy in Brazil and finishing with an approach of the Brazilian bioclimatic zones with very different prevalent climates. In this section it is also presented the Brazilian legislation associated to the study of the thermal performance of buildings built in the country: NBR 15220-3 [5] concerning the different bioclimatic zones and NBR 15575-5 [6] which refers to the thermal performance of single-family residential buildings and multi-familiars, in addition to RTQ-R regulation whose function is to complement the two last standards with regard to the issue of energy efficiency in buildings;

- Chapter "Experimental Analyse of the Influence of Different Mortar Rendering Layers in Masonry Buildings" presents a brief analyses of the thermal comfort, comfort-related concepts and their evaluation parameters, presents the models proposed by the standards comfort ASHRAE 55 [7], EN 15251 [8] and ISO 7730 [9], deals with the LNEC model and its similarities with ASHRAE 55 model and culminates with the presentation and justification of the model adopted to delineate the comfort in this research works. In this section is still exposed a study on night ventilation and their contributions to the comfort as well as a synthesis of the studies related to the topic;

- Finally, Chapter "Ceramic Membranes: Theory and Engineering Applications" presents an approach related to the plaster; make an exploratory study about advancing this constructive modality, particularly in the north-eastern region of Brazil and their projections to the country. It is presented a national and international history of legislation regarding the use of plaster as constructive material. Some questions addressed in this section, such as durability, has not been analysed in this work and the authors suggest their study in future researches.

2 Brazilian Housing Deficit and Thermal Performance Requirements

2.1 Brazilian Housing Deficit

Maricato [10] states that a significant number of Brazilians lives in housing precarious conditions. A research of housing deficit in Brazil, developed by the João Pinheiro Foundation (FJP) [3], in partnership with Minister of Cities, estimated that, in 2007, the housing deficit in Brazil is approximately of 6,300,000 of households; and more than 10,500,000 households were not equipped with any infrastructure. These results show that about 40–50% of the population of the large Brazilian centers living in precarious conditions, being 20% in slums.

A study published by the Minister of Cities-MC [11], at the same time, ensures that 92% of the Brazilian housing deficit corresponds to a population with income between zero and 5 minimum salaries; and 84% of this value, refers to Brazilians framed income range between 0 and 3 minimum wages.

Data prepared by João Pinheiro Foundation (FJP) [12] based on statistical developed by IBGE/PNAD-2012, in Brazil, showed that despite the advances of the growth acceleration program (PAC) of the Federal Government through the program "My house my life" (PMCMV), it is possible to observe a housing deficit of about 5,400,000, distributed among 4,600,000 urban households and approximately 800,000 rural houses. In 2013, the most recent data published by the FJP and updated by the Brazilian construction industry (CBIC) in conjunction with the IBGE/PNAD, showed a deficit of 5,800,000 dwellings, being this amount divided for 5 million urban and 800,000 unit of rural housing. Therefore, it is possible to observe, between the years 2012 and 2013, an increase of 7.4% in the Brazilian housing deficit.

According the IBGE data (Census 2010 [1]), there are currently in Brazil about 57,300,000 households considered private and permanent. Ten years ago, according to the same source, these households were 44,700,000, registering an increase of about 28% [1].

In this same period the Brazilian population grew from approximately 170 million to 191 million (see Table 1), representing this period an increase of 12.3% of people, which amounts to less than half of registered increase of housing units available in the country.

According to IBGE results [1], Sao Paulo, the most populous Brazilian city, the vertical construction (apartments) holds to 12,800,000. The last official data from the IBGE [1] concerning the type of Brazilian housing showed an increase of 43% in the number of apartments in the last decade (2000–2010), from 4,300,000 for 6,100,000. Approximately half of this value is located in the Brazilian southeast region, distributed by 1,800,000 in Sao Paulo and 1 million in Rio de Janeiro. The Tocantins state has the lowest concentration of this type of property, adding only 5447 apartments.

The buildings built in blocks of plaster were not accounted in the statistics data presented by IBGE/PNAD, because this type of construction doesn't present a significant number. Garcia and Castle [13] consider the housing *deficit* not only the lack of housing, but those that are uninhabitable.

In Brazil, according to the FJP, millions of households are excluded from access to decent housing, which corresponds to a deficit of 7,200,000 new dwellings. Of these values, 5,500,000 are in urban areas and 1,700,000 in rural areas. Southeast

Table 1 Population and Relative distribution (%) for Brazil and its major regions [1]

Brazil and its major regions	Population		Relative distribution	
	2000	2010	2000	2010
Brazil	169,799,170	190,755,799	100.0	100.0
North	12,900,704	15,864,454	7.6	8.3
Northeast	47,741,711	53,081,950	28.1	27.8
Southeast	71,421,411	80,364,410	42.6	42.1
South	25,107,616	27,386,891	14.8	14.4
Midwest	11,636,728	14,058,094	6.9	7.4

Table 2 Brazilian housing deficit [3]

Brazil and its major regions	Housing shortage			Relative percentage of households			Absolute percentage of households		
	Total	Urban	Rural	Total	Urban	Rural	Total	Urban	Rural
Brazil	7,222,645	5,469,851	1,752,794	16.1	14.6	23.7	100	100	100
North	848,696	506,671	342,025	30.2	24.8	44.6	11.8	9.3	19.5
Northeast	2,851,197	1,811,553	1,039,644	25.0	22.2	32.1	39.5	33.1	59.3
Southeast	2,341,698	2,162,187	179,511	11.6	11.7	10.4	32.4	39.5	10.2
South	678,879	565,217	113,662	9.4	9.5	8.9	9.4	10.3	6.5
Midwest	502,175	424,223	77,952	15.9	15.4	19.3	6.9	7.8	4.5

(32.4%) and Northeast (39.5%) are the regions that represent the highest housing demands of the country (see Table 2).

There is a predominance of housing deficit, both quantitative and qualitative, concentrate in urban areas and in the lower income ranges of the population, mainly in the metropolitan areas.

In 2000, 91% of the country's housing deficit had correspondence with the families whose financial income reached up to five minimum wages.

Data from the João Pinheiro Foundation [12], present in Table 3, show that 83% of the Brazilian housing deficit corresponds to families who earn up to three minimum wages. The largest absolute quantitative income on this issue highlights the north-eastern and South-eastern regions with 91.3 and 77.6%, respectively.

Denaldi [14], in a brief retrospective of as is treated the issue of housing policies in the country, showed the absence of the Brazilian government in the planning of the sector, even at the time of the intervention, and that it had not been possible to meet the needs of lower income population, especially the population with incomes that not exceed three minimum wages. According to the author, the urbanistic

Table 3 Housing shortage by family income [12]

Territory	Monthly income in minimum wages							
	Up to 3 MW[1]		>3–5		>5–10		>10	
	N°	%	N°	%	N°	%	N°	%
North	343,301	84	29,235	7.2	28,258	6.9	6456	1.6
Northeast	1,554,079	91.3	87,333	5.1	35,963	2.1	11,604	0.7
Pernambuco	279,823	91.0	14,525	4.7	6871	2.2	2066	0.7
MR[2] Recife	164,652	88.3	9585	5.1	5322	2.9	1039	0.6
Southeast	1,694,803	77.6	239,257	11.0	154,648	7.1	64,613	3.0
South	465,063	80.9	54,020	9.4	38,404	6.7	14,286	2.5
Midwest	353,139	82.9	33,294	7.8	27,858	6.5	8,673	2.0
Brazil	4,410,385	83.2	443,139	8.4	285,131	5.4	101,632	1.9

MW[1]—minimum wage
MR[2]—metropolitan region

instruments were applied in order to meet the interests of an elite and the real estate market, generating an increase in social inequalities.

After 2003, the Minister of Cities (MC) was in charge of the function of coordination, management and formulation of the National Urban Development Policy, where it was included in this, the new National Housing Policy (PNH), whose approval, in November 2004, established a new institutional organization model based on National Housing System (SNH).

Since then, the National Housing System went to become the main instrument of National Housing Policy, providing the integration of the Brazilian government with public and private actors involved directly in the subject of the housing issue.

By the 2030, according E&Y [15], the country should aim for solving the housing deficit; take into account the emergence of new families. Also according E&Y [15], in 2030 the country will have an estimated contingent over 233 million people and about 95,500,000 families. A statistical analyse shown that there will be an estimated average of 2.5 persons per dwelling, which will culminate in about 93,100,000 households causing a growth of approximately 66% in comparison to 2007. During this period, should be constructing 37 million of new house across the country, which could mitigate the housing deficit with an average of 1.6 million new homes a year.

It is clear that the accumulated Brazilian housing deficit is a challenge to be faced with correct housing policies and previously established rules. According to estimates developed by Joao Pinheiro Foundation of Minas Gerais and based on information collected by PNAD/IBGE [16], the Brazilian housing deficit, this year, is the approximately 7,900,000 new housing units. According to projections made by CEDEPLAR [17] and adopted by the PLANHAB (Brazilian National Housing Plan), for the horizon 2007–2023 that demand would still be of the order of 27 million units, taking into account the projection of formation of each family in the estimated period.

The results of the study of demands made by CEDEPLAR [17] can be appreciated through the analysis of Table 4 and Fig. 1, with the breadth of the municipalities involved and the typology adopted (A–K). Still in Table 4 it is possible to check the framing shape of the cities and metropolitan regions, as well as briefly the economic and methodological criteria used in the selection of the Brazilian municipalities to be served at the time of planning. The absolute values provided for in the study are presented in Table 5, for the years of 2007–2023.

A more critical analysis of Table 5 and Fig. 1, showed that the municipalities covered in the study prepared by CEDEPLAR/UFMG [17], was more to the municipalities included in types of classes I, J and K, corresponding to those with population of less than 20,000 inhabitants, with predominant rural characteristics and low density. This criterion covers most of the municipalities located in the Northeast and North regions of Brazil, and some municipalities of the southern and south-eastern Brazil.

In resume, according this estimate, the construction of about 35 million of new houses should be sufficient to mitigate the Brazilian housing deficit during the next 15 years (Table 6).

Table 4 Classification of urban municipality typology and metropolitan region on demand [17]

Municipalities	Typo	Regions involved and criteria
Municipalities of metropolitan areas and municipalities with more than 100 thousand inhabitants	A	Metropolitan area of Rio de Janeiro and São Paulo. Cities located in high-income regions, with high social inequality. Are called global metropolis by the concentration of economic and financial activities
	B	Metropolitan regions and major agglomerations of the Centre-South Urban agglomerations and capital cities located in regions of high stock of wealth. Great functional importance in the network of cities. Are considered cities poles in their respective regions
	C	Metropolitan regions and major agglomerations and prosperous capitals of the N and NE. Main centres in the North and Northeast, polarizers with wealth below the stock types A and B, with the highest concentration of poverty and high inequality
	D	Clusters and regional centres of the Centre-South. Municipalities located in regions with high stock of wealth, with importance as polarizing centres in your region
	E	Clusters and regional centres N and NE. Municipalities with low stock of wealth, but with great regional importance. Cities located in Brazil with lower poles dynamism
Municipalities with population between 20 and 100 thousand inhabitants	F	Cities in prosperous rural areas. Municipalities that are growing moderately, situated in the country's richest regions. Have more policies to tackle the deficit with own resources
	G	Urban centres in rural areas consolidated with some degree of dynamism Municipalities located in micro-regions historically greater poverty and relative stagnation, but exhibit more positive situation compared to other types
	H	Urban centres in rural areas with high inequality and poverty. Municipalities that stand out by higher levels of poverty, a greater number of households with no bathroom, and high housing deficit

(continued)

Table 4 (continued)

Municipalities	Typo	Regions involved and criteria
Municipalities with population of less than 20 thousand inhabitants	I	Small towns in rural areas
	J	Small towns in rural areas consolidated, but fragile recent dynamism
	K	Small towns in rural areas with low economic density

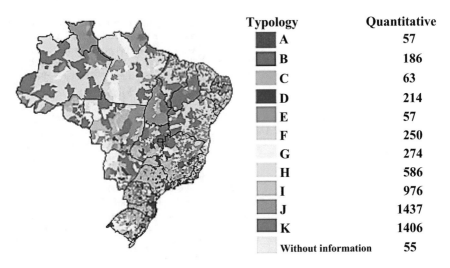

Typology		Quantitative
A		57
B		186
C		63
D		214
E		57
F		250
G		274
H		586
I		976
J		1437
K		1406
Without information		55

Fig. 1 Demand in Brazil by typology [17]

Table 5 Projection of future new houses for 2007–2023 [17]

Projection							
Specification	Accumulated deficit	2007–2011	2012–2015	2015–2019	2020–2023	2007–2023	Total
New units missing	7.90	8.73	5.20	5.86	7.19	26.98	34.9
Households in precarious conditions	3.26						
Households that needed complementary infrastructure	9.83						

According to the TIS (Territorial Information Survey) carried out on the basis of the Census (IBGE) [1], the Southeast region presents almost half of the households in poor communities of the country. According to the survey conducted from data collected in 323 Brazilian municipalities with records of the existence of slums, Rio de Janeiro, São Paulo, Minas Gerais and Espirito Santo concentrate 49.8% of total slums in Brazil.

Table 6 Housing needs scenario for Brazil (2007–2023)

Municipalities		Households needs scenario				Total (%)	
		2007–2011	2012–2015	2016–2019	2020–2023		
Urban	A	1,235,178	714,974	815,067	1,044,013	3,809,231	14.1
	B	945,998	545,023	613,798	757,517	2,862,335	10.6
	C	617,374	363,949	402,550	500,603	1,884,476	7.0
	D	1,066,716	647,237	738,116	897,491	3,349,560	12.4
	E	631,529	394,835	438,514	524,028	1,988,907	7.4
	F	453,464	270,167	306,664	380,763	1,411,059	5.2
	G	535,381	323,977	364,458	438,356	1,662,172	6.2
	H	577,029	344,543	383,070	466,415	1,771,057	6.6
	I	240,127	144,579	163,823	201,015	749,544	2.8
	J	346,961	211,118	236,347	283,204	1,077,630	4.0
	K	324,031	192,796	215,272	259,717	991,815	3.7
Rural		1,757,171	1,056,528	1,184,785	1,432,083	5,430,567	20.1
Total		8,730,960	5,209,726	5,862,462	7,185,205	26,988,353	100.0
		32.4%	19.3%	21.7%	26.6%	100.0%	

Source PNAD/IBGE [16]

Related to the 3.2 millions of households in subnormal clusters, called as slums by IBGE [3], and dispersed by 27 Brazilian regions, 1.6 million are in the southeast, which is the most populous region of the country. The Brazilian northeast, the second region with largest numbers of homes in poor communities, counts with 28.5% of slums which equates to 926,000 households. Then, the north region, with 14.4% slums which equates to 463,000 households; the south of the country, with 5.3% slums which equates to 57,000 units, and finally, the Midwest region, which counts only 1.8%, corresponding to an aggregation of 170,000 dwellings (see Fig. 2). According to the survey of Census 2010 [1] the number of Brazilians that lives in slums is greater than 11.4 million.

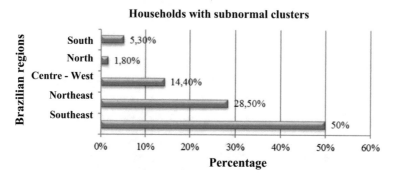

Fig. 2 Distribution of households with subnormal clusters in Brazil [1]

The Brazilian southeast region, with 50% of buildings considered subnormal clusters (see Fig. 2), is the most populous regions of the country with 80.3 million of inhabitants according to the IBGE census 2010 [1], with a territorial area of 924,500 Km^2 and a population density of 87 inhabitants per Km^2. The North and Midwest presents the smaller percentage (1.8%) of subnormal clusters, and smallest population in the country, approx. 14.9 million inhabitants and 9.4 inhabitants per km^2. In conclusion, according to the IBGE census 2010 [1], 11.4 million of Brazilians lives in slums.

In March 2009, the Brazilian government through the Ministry of cities (MC) based on studies of the FJP and the HNP, created the Project My House My Life (PMCMV). According to the Law 11,977 [18], was deployed the PMCMV program, whose purpose was the creation of incentive mechanisms to the production and acquisition of new housing units by families with a salary less than 10 minimum wages.

In quantitative terms, the PMCMV, initially, foresaw the construction of one million dwellings distributed according to the population monthly salary range, including 400,000 families with income of up to three minimum wages, 400,000 with income between three and six minimum wages and 200,000 with a salary range between six and ten minimum wages. The initial numbers of the Law n° 11977 [18], would not solve the Brazilian housing deficit, however, constituted one of the first housing programs offered by the federal government in partnership with the Minister of Cities, whose operational management competed to *Caixa Económica Federal* bank, who makes this bank an important partner with regard to housing policies in Brazil. Figure 3 comes briefly illustrate the investments granted by the government for housing, until 2015, and demonstrates a period of stagnation with little growth of investments in the sector prior to the enactment of the Law 11977 [18], in 2009.

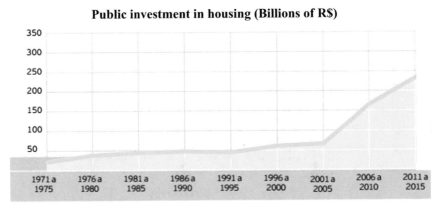

Public investment in housing (Billions of R$)

Fig. 3 Brazilian allowances for housing policy. Adapted from [19]

In resume, the financial resources that initially were targeted to the urbanization of slums had a significant increase with the creation of the governmental program of acceleration of growth (PAC), which has earmarked about R$11 thousand million in the period of 2007–2010. The priority of this Government program included the metropolitan areas of major cities, urban agglomerations and cities with more than 150 thousand inhabitants.

2.2 Brazilian Energy Issue

Since the oil crisis caused by the reduction in supply in the 70's, practically the debating the energy issue is discussed in the entire world, mainly in developed countries. Brazil, on the other hand, has maintained a favourable position in relation to other countries, since your energy matrix is in great part based on renewable sources [20].

However, despite the large energy matrix, in the middle of 80s, Brazil presented some problems related to the energy crisis caused by great demand, which exceeded the growth in energy production. In the 80s, began the first actions related to the theme of energy efficiency in the country. It was created the PROCEL program (national program for conservation of electrical energy), developed by the Ministry of Mines and Energy (MME), in 1985, which the main objective was the promotion policy and expansion of energy production. Since then, PROCEL has advanced in the research of energy efficiency in all the country, promoting comprehensively the establishment of research partnerships with different institutions [21].

Despite this numerous efforts, in 2001 Brazil had an energy crisis which forced the population to meet a target of 20% electricity consumption reduction, caused by a bad sector planning. In 2001, Brazilian government created the Law No. 10295 [22], that established the maximum levels of specific energy consumption for electric machinery manufactured in the country. However, in the opposite direction to the law, the changes in the form of life and the increase of the population of the cities, with the acquisition of new consumer goods and the emergence of the electricity grid in places where it did not exist before, contributed to the increase the energy consumption, despite Government efforts to encourage low consumption. Figure 4 shows a comparison between energy consumption and population growth in the period between 1990 and 2010.

Energy plays a key role in a modern society, acts as a key element in social issue by providing that a sustainable development and in improving the quality of life. [22]. A methodical research for a better comfort conditions in buildings should contributes to the reduction of electric power supply consumption, especially in developing countries. It is necessary to reduce the presence, in large-scale, of mechanisms for artificial cooling that increase the consumption of electric energy, unnecessarily.

Huberman and Pearlmutter [23] show that the world demands for energy with a trend of growth that can achieve rates of 71% between 2003 and 2030, motivated by the thermal comfort in buildings. A study developed by Kuznik et al. [24]

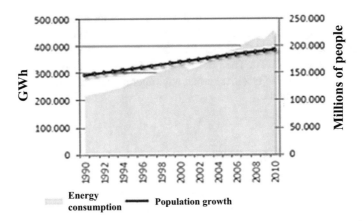

Fig. 4 Population growth versus energy consumption increases—Period 1990–2010 [4]

show an energy growth in the order of 30% over the last 30 years only in buildings. According to Achão [25], despite the energy growth rate in Brazil, there are problems of distribution by region and social class, which implying directly on unequal offer comfort.

In Brazil, in recent years, have developed relevant studies related to the theme of energy efficiency in buildings. These researches are being encouraged by the federal Government, in response to the Law 10295 [22], in your article 4, through the National Program of Combating Electrical Waste (PROCEL), with the participation of other entities. Already exists in the country a reliable diagnosis of the main problems, actions required and centers of excellence with experts on these specific themes.

In the data analysis, by energy sector consumption, provided by BEN-National Energy Balance [20], it is possible to observe a predominance of energy consumption by the industrial sector, with 44% of the total energy supply, in accordance with the good industrial development phase of Brazil. Adding the energy consummation of the residential, commercial and public sectors, a value of 47% was obtained, causing these sectors a great concern, particularly the residential sector (see Fig. 5).

In conclusion, a great importance should be given to the residential sector, responsible for 24% of electricity consumption, in the country, in 2009, and with an increasing rate of 5.7% per year, for the records during the period of 1975–2009 [20]. Among the factors that contributed to this increase, it is possible to highlight:

- Increased income, particularly for the low-income classes;
- Substantial households growth;
- Increase of new household goods that consume electric power, in particular by the middle and poor class;
- An increase of the lines of credit;
- Creation of small informal businesses, in residences.

Fig. 5 Electric Power
Consumption, by sector, in
Brazil [20]

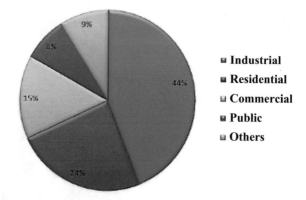

■ **Industrial**
■ **Residential**
■ **Commercial**
■ **Public**
■ **Others**

Brazil's energy matrix presents a predominance of power generation from natural sources, however, in recent years, due to unfavourable hydrological conditions, there was a substantial reduction in the supply of hydropower (in 2014 this reduction was 5.5% compared to 2013). In this same period (2013–2014), it was records a growth of wind energy of approximately 1%. In 2014 the wind energy represented an increase of 12,100 GWh in total production, with 167 wind farms installed in the country. In 2015, it went into operation more than 100 wind power plants in Brazil and it was recorded a growth of 57% in that year. Brazil is, currently, the fourth country, with a higher growth in wind energy.

In conclusion, the rapid and consistent growth of Brazilian population and consequently the increase of energy demand, justify the research for new construction methodologies, with new materials, as proposed in this work.

2.3 Brazilian Bioclimatic Zones

The Brazilian bioclimatic zones are part of the NBR 15220-3 [5], since 2005. Brazilian territory is divided in 8 climatic zones. For each zone, the standard makes recommendations of passive thermal conditioning strategies for social housing schemes. The goal of the technical-constructive regulatory recommendations is the optimization of the thermal performance of buildings through better fitness. Environmental conditioning strategies recommended by the same standard are based on the bioclimatic chart of Givoni [26] and in Mahoney worksheets [27].

The NBR 15220-3 [5] divides the country in eight bioclimatic zones, being considered in this work the abbreviations Z1 to Z8 for each of the zones. Bioclimatic zone 1 (Z1) is the coldest and also the smallest, representation an area of 0.8% only. Zone 8 (Z8) is the hottest and the biggest bioclimatic zone in terms of covered area (57%), it includes the North and Northeast, and a large portion of the Brazilian coast (see Fig. 6).

Fig. 6 Brazilian Bioclimatic
zones [27]

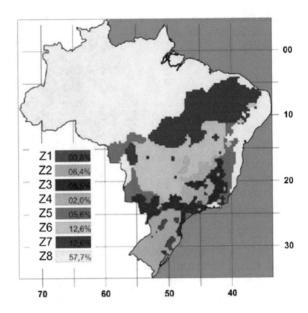

2.3.1 Bioclimatic Zone 1 [Z1]

The Z1 includes a total of 12 Brazilian cities, with a strong predominance of the cities located in the Brazilian southern region. Examples are the cities of Caxias do Sul (RS), Lages (SC), Curitiba (PR) and Poços de Caldas (MG) located in the Southeast. For this Bioclimatic Zone, the NBR 15220-3 [5] recommends as a strategy of passive thermal conditioning, heavy internal seals with greater thermal inertia and solar heating of buildings in winter season. Figure 7 illustrates the outdoor temperature and relative humidity in 2016 for the city of Curitiba (PR), as example.

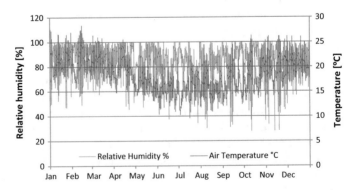

Fig. 7 Temperature and relative humidity variation for the city of Curitiba [28]

2.3.2 Bioclimatic Zone 2 [Z2]

This bioclimatic zone comprises 33 cities that make up the greater prevalence of these Z2 in the southern region of the country; however, there are a small number of cities, with the same climatic characteristics, in the Southeast region of Brazil. Among some cities included in this area, are the cases of Alegrete (RS) and Nova Friburgo (RJ). The Brazilian standard, NBR 15220-3 [5], indicates conditioning strategies for the summer and winter seasons and the use of cross-ventilation in summer suggests the winter solar heating and indicates internal seals for this season. To this bioclimatic Zone Z2, NBR-15220 indicates average openings allowing the entry of sunlight during the winter season [28]. Figure 8 illustrates the outdoor temperature and relative humidity in 2016 for the city of Passo Fundo (RS).

2.3.3 Bioclimatic Zone 3 [Z3]

This bioclimatic zone includes 62 cities, among which is São Paulo, the city with the largest concentration of population in the country.

The NBR-15220-3 suggests the adoption of cross-ventilation for the summer period as a strategy of passive thermal conditioning, the solar heating for the winter season, and a strategy for the external seals, light walls and light covers [5]. Openings must be medium and allow the entry of sunshine during the winter. Figure 9 illustrates the climatic data of temperature/relative humidity for the city of São Paulo.

2.3.4 Bioclimatic Zone 4 [Z4]

NBR 15220-3 [5] establishes as a strategy of passive thermal conditioning, evaporative cooling and ventilation, for the summer period, and solar heating of

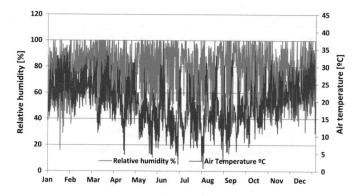

Fig. 8 Temperature and relative humidity variation for the city of Passo Fundo [28]

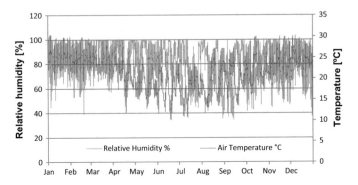

Fig. 9 Temperature and relative humidity variation for the city of São Paulo [28]

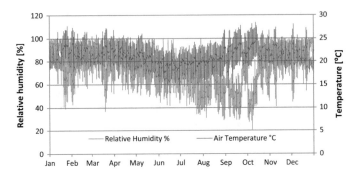

Fig. 10 Temperature and relative humidity variation for the city of Brasília [28]

buildings, as well as internal seals, i.e. greater requirement for the thermal inertia in buildings, for winter. In relation to external seals, the recommendation of the standard indicates massive walls. However, it is recommended that the coverage can be mild, however, must be isolated. For the city of Brasília, the capital of the country situated in this zone 4, is verified through bioclimatic design of Fig. 10, a low relative humidity in almost all year and minimum critical values between the months of August to October, with indexes under 40%.

2.3.5 Bioclimatic Zone 5 [Z5]

The NBR 15220-3 [5] recommends for this bioclimatic zone the use of cross-ventilation in summer season and heavy internal seals for the winter season as passive thermal conditioning strategies in this bioclimatic area. Ventilation openings in this area should be medium, however, the standard indicates the shading of these openings. Are included in this zone 5, in the Northeast, the cities of Garanhuns-PE, Triunfo-PE, Morro do Chapéu-BA and Guaramiranga-CE. In the

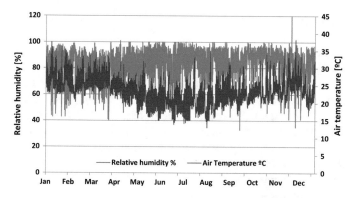

Fig. 11 Temperature and relative humidity variation for the city of Niteroi [28]

Southeast, the cities of Niterói-RJ, Governador Valadares and Pedra Azul in the State of Minas Gerais. Figure 11 illustrates the climatic data of temperature/relative humidity for the city of Niterói, Rio de Janeiro State.

2.3.6 Bioclimatic Zone 6 [Z6]

NBR 15220-3 [5] recommends, for Z6, the use of selective ventilation (for the periods in which the internal temperature is higher than the outdoor temperature), which checks for elevated temperatures throughout much of the year. The seals, according to the normative recommendations, should be with heavy walls, the coverage can be light, however, isolated. Openings must be medium sized and shaded. Figure 12 illustrates the climatic data of temperature/relative humidity for the city of Bom Jesus da Lapa, Bahia.

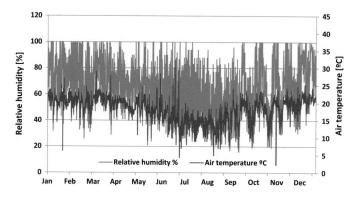

Fig. 12 Temperature and relative humidity variation for the city of Bom Jesus da Lapa [28]

2.3.7 Bioclimatic Zone 7 [Z7]

According to NBR 15220-3 [5], should prioritized projects involving buildings with shaded small apertures for ventilation, the external cover seals and walls must be heavy, and must adopt thermal conditioning strategies through evaporative cooling and ventilation must be selective, whenever the internal temperature is higher than the outside. The evaporative cooling strategy has the ability to reduce the temperature and increase the relative humidity of a room. Although of little use in Brazil, evaporative cooling systems have enormous potential to provide thermal comfort in places where the wet bulb temperature is relatively low. Data of temperature and relative humidity of for the city of Petrolina, Pernambuco, place of study of the prototype, are presented in Fig. 13.

2.3.8 Bioclimatic Zone 8 [Z8]

According to NBR 15220-3 [5], the buildings must contain shaded large openings for ventilation and adopt cross-ventilation strategies permanently in the summer period. However, the Brazilian standard does not prescribe any type of strategy of conditioning for the winter season. Temperature and relative humidity data for the city of Recife-PE is presented in Fig. 14.

Figure 15 presents a summary of cumulative probability curves that representing the 7 Brazilian bioclimatic zones. It is possible to observe, in more detail, the temperature variation between the predominant climates of Brazil.

According to NBR 15220-3 [5], these guidelines were developed in order to guide the buildings to adapt and take advantage of local climate conditions. The conditions of comfort and energy consumption are dependent variables, which, if a building with their constructive characteristics is not adapted to the local climatic conditions, the comfort conditions are not taken into account and consequently results in an increase of energy consumption.

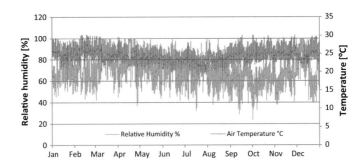

Fig. 13 Temperature and relative humidity variation for the city of Petrolina [28]

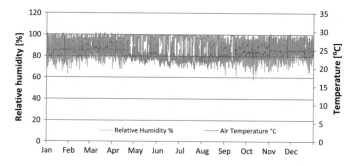

Fig. 14 Temperature and relative humidity variation for the city of Recife [28]

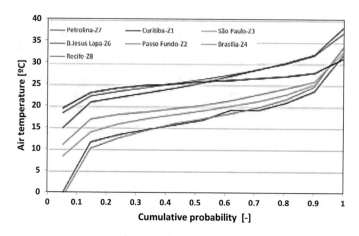

Fig. 15 Cumulative probability curves for the 7 bioclimatic zones in Brazil

In this section it was possible to present an explanation of the Brazilian Bioclimatic with the respective zoning areas and percentage of coverage of each zone. Outside air temperatures and relative humidity were observed in eight cities that representing each of these bioclimatic zones. In this study, it was possible to identify two bioclimatic zones with negative temperatures, Curitiba in Z1 and Passo Fundo in Z2. Two cities, Petrolina in Z7 and Bom Jesus da Lapa in Z6 present very similar temperature curves although they are in different locations. This analysis may indicate a possible failure in the framework criteria of NBR 15220-3 [5], for some cities located in different bioclimatic zones, but with similar climatic characteristics.

2.4 Buildings Thermal Performance Regulation in Brazil

2.4.1 NBR 15220 (2005)—Thermal Performance of Buildings

This Brazilian standard for buildings thermal performance was published in 2005, and it is divided in 5 parts, the first part presents the definitions and symbols, the second presents the calculation methodology of the elements and components of buildings. On part 3 it was described the Brazilian Bioclimatic Zones, construction guidelines and thermal conditioning strategies for each zone. Parts 4 and 5 presents the method of measurement of thermal resistance and thermal conductivity by hot plate principle.

NBR 15220-3 [5] presents the guidelines for the use of ventilation in seven bioclimatic zones, not recommending ventilation only in zone 1, because it is the coldest region of the country. Cross-ventilation was recommended for the bioclimatic zones 2, 3 and 5, where it is possible to maintain the air flow through the vents, doors and windows. The ventilation setting, recommended for zones 4, 6 and 7, are indicated in warmer periods or when the internal temperature is higher than the external; and permanent ventilation is recommended for bioclimatic zone 8, and it is justifying by the necessity to improve the indoor thermal conditions of the buildings.

2.4.2 NBR 15575 (2013)—Residential Buildings Performance

NBR 15575 [29] standard, in your fourth edition, was published in 2013 and present the conditions of habitability for the range of environmental comfort, through the minimum conditions of thermal comfort, luminosity, hygrothermal, visual and psychological for the users. It is divided into six parts, as the following:

- Part 1—general requirements;
- Part 2—requirements for the structural systems;
- Part 3—requirements for floor systems;
- Part 4—requirements for the systems of internal and external seals;
- Part 5—requirements for the systems of coverage;
- Part 6—requirements for hydro-sanitary systems.

NBR 15575 [30] presents two procedures to evaluate the thermal performance of buildings; the first procedure analyse the construction in design phase through numerical simulation, and the other procedure evaluates the building already executed, through in situ measurements. For the evaluation of thermal performance by numerical simulation, the standard presents three performance levels (minimum, intermediate and higher level) for the building, relating the interior temperature with the temperature outside in the winter and summer seasons, according to the bioclimatic zone in which the building is inserted (see Table 7).

Table 7 Thermal performance evaluation criteria for winter and summer

NBR 15575 performance levels

Winter criterion				Summer criterion		
Level	Temperature	Application to bioclimatic zone		Level	Zone Z1 to Z7	Z8
		Z1 to Z5	Z6 to Z8 does not check these criteria			
M	$T_{i,min} \geq (T_{o,min} + 3\ ^\circ C)$			M	$T_{i,max} \leq T_{o,max}$	$T_{i,max} \leq T_{o,max}$
I	$T_{i,min} \geq (T_{o,min} + 5\ ^\circ C)$			I	$T_{i,max} \leq (T_{o,max} - 2\ ^\circ C)$	$T_{i,max} \leq (T_{o,max} - 1\ ^\circ C)$
S	$T_{i,min} \geq (T_{o,min} + 7\ ^\circ C)$			S	$T_{i,max} \leq (T_{o,max} - 4\ ^\circ C)$	$T_{i,max} \leq (T_{o,max} - 2\ ^\circ C)$ and $T_{i,min} \leq (T_{o,min} + 1\ ^\circ C)$

Adapted from NBR 15575 [5, 6]

Performance levels: M—minimum, I—intermediate, S—maximum

$T_{i,max}$—daily maximum indoor temperature in °C

$T_{i,min}$—daily minimum indoor temperature in °C

$T_{o,max}$—daily maximum outdoor temperature in °C

$T_{o,min}$—daily minimum outdoor temperature in °C

In summary, Table 7 shows that it is not necessary to verify the level of thermal performance of a building located in Z6 and Z8, in winter season, but only for the summer season. However, these regions present different climatic conditions, for example, Bom Jesus da Lapa, located in Z6, with an altitude above 400 m, and 800 km away from the coast and the city of Recife, located in Z8, on the coast of Pernambuco, with an altitude of 4 m, with more mild temperatures that Bom Jesus da Lapa, but both subject to the same conditions of winter analysis. These premises influence the need for an urgent review of the standards.

2.4.3 RTQ-R (2012)—Quality Technical Regulation of the Energy Efficiency Level of Single-Family and Multi-family Residential Buildings

RTQ-R [30], presented in 2010, specifies the technical requirements and test methods for the energy efficiency classification of buildings. This standard presents the labelling of buildings and analyses the building envelope requirements, related to lighting system, air-conditioning system and the construction as a whole. Establishes guidelines for thermal comfort considering the vent through a minimum opening area in long-stay environments, as bedroom, living room and kitchen.

Among other aspects, in RTQ-R are defined the actions related to the range of passive thermal condition, cross-ventilation, ventilation and other control devices that determine the minimum level of comfort and thermal performance of buildings. The regulation considers as mechanical devices, the blinds, ventilated sills (air intakes which affect cross-ventilation), the wind towers (extraction ducts), as well as external openings through windows.

In relation to the openings and natural ventilation, the RTQ-R [30] provides the minimum percentages of ventilation areas for each bioclimatic zone and determines the cross-ventilation for seven of the eight bioclimatic zones of Brazil.

NBR 15220-3 [5] considers three distinct dimensions of openings for ventilation in buildings: small, medium and large openings, specifying the percentages of these openings by framing them with the area of the floor. NBR 15575-4 [29] specifies the minimum percentage values of these gaps related to the area of the floor, enforcing your applicability to the eight bioclimatic zones according to the standard.

Table 8 Percentage of opening ventilation limits [5, 30]

NBR 15220-3		NBR 15575-4		
Opening ventilation	A (% floor area)	Zone	Locations	A (% floor area)
Small	10% < A < 15%	Z1–Z6	Extended-stay environments and Kitchen	A $\geq$ 8%
Medium	15% < A 25%	Z7		A $\geq$ 5%
Large	A > 40%	Z8		A $\geq$ 10%

Table 9 Surrounding thermal performance indicators [5, 6]

Zone	External walls NBR 15220-3			NBR 15575-5		Roofs NBR 15220-3			NBR 15575-5	
	U	φ	FS_0	U_{min}	CT min	U	φ	FS_0	U	α
1	≤ 3.00 (light)	≤ 4.3	≤ 5.0	≤ 2.5 (light)	≥ 130	≤ 2.00 (isolated)	≤ 3.3	≤ 6.5	≤ 2.30	–
2	≤ 3.00 (light)	≤ 4.3	≤ 5.0	≤ 2.5 (light)	≥ 130	≤ 2.00 (isolated)	≤ 3.3	≤ 6.5	≤ 2.30	–
3	≤ 3.60 (reflective light)	≤ 4.3	≤ 4.0	≤ 3.7 and $\alpha \leq 0.60$	≥ 130	≤ 2.00 (isolated)	≤ 3.3	≤ 6.5	≤ 2.30	$\alpha \leq 0.6$
				≤ 2.5 and $\alpha > 0.60$					≤ 1.50	$\alpha > 0.6$
4	≤ 2.20 (heavy)	≥ 6.5	≤ 3.5	≤ 3.7 and $\alpha \leq 0.60$	≥ 130	≤ 2.00 (isolated)	≤ 3.3	≤ 6.5	≤ 2.30	$\alpha \leq 0.6$
				≤ 2.5 and $\alpha > 0.60$					≤ 1.50	$\alpha > 0.6$
5	≤ 3.60 (reflective light)	≤ 4.3	≤ 4.0	≤ 3.7 and $\alpha \leq 0.60$	≥ 130	≤ 2.00 (isolated)	≤ 3.3	≤ 6.5	≤ 2.30	$\alpha \leq 0.6$
				≤ 2.5 and $\alpha > 0.60$					≤ 1.50	$\alpha > 0.6$
6	≤ 2.20 (heavy)	≥ 6.5	≤ 3.5	≤ 3.7 and $\alpha \leq 0.60$	≥ 130	≤ 2.00 (isolated)	≤ 3.3	≤ 6.5	≤ 2.30	$\alpha \leq 0.6$
				≤ 2.5 and $\alpha > 0.60$					≤ 1.50	$\alpha > 0.6$
7	≤ 2.20 (heavy)	≥ 6.5	≤ 3.5	≤ 3.7 and $\alpha \leq 0.60$	≥ 130	≤ 2.00 (isolated)	≤ 3.3	≤ 6.5	≤ 2.30 FV	$\alpha \leq 0.4$
				≤ 2.5 and $\alpha > 0.60$					≤ 1.50 FV	$\alpha > 0.4$
8	≤ 3.60 (reflective light)	≤ 4.3	≤ 4.0	≤ 3.7 and $\alpha \leq 0.60$	No requirement	≤ 2.00 (isolated)	≤ 3.3	≤ 6.5	≤ 2.30 FV	$\alpha \leq 0.4$
				≤ 2.5 and $\alpha > 0.60$					≤ 1.50 FV	$\alpha > 0.4$

FV—ventilation factor established by NBR 15220-2

FT—correction factor that allows higher transmission in penthouses with ventilated attics

U—thermal transmission coefficients (W/m^2 °C)

φ—thermal delay component (h)

α—solar radiation absorptivity of outer surface of the cover

CT—heat capacity (kJ/m^2 K)

FSo = g$\perp$—solar factor (%)

For example, Petrolina located in Z7, in accordance with NBR 15220-3 [5], should have buildings with an average openings ranging between 15 and 25% of the floor area, however, in accordance with NBR 15575-4 [29], these openings should be equal to or greater than 5%. The recommended percentages of these 2 standards may be seen in Table 8.

In summary, for the analysis of Table 8, it is possible to observe a greater flexibility of NBR 15575-4 [29] with respect to NBR 15220-3 [5] standard. The thermal performance of surrounding buildings projects is presented in Table 9, with a comparison between the standards NBR 15220-3 [5] and NBR 15575-5 [6] which establishes the maximum residue limits and the minimum values of thermal transmittance, thermal and solar factor delay for each case of surrounding and its applications to the different bioclimatic zones.

For a better understanding of the thermal parameters presented in Table 9, it is appropriate to clarify that the Ventilation Factor (FV) is a parameter defined in the Brazilian standards NBR 15575, NBR 15220-3 and NBR 15220-5 and it's calculated in function of the size of vents in eaves, given by the expression:

$$FV = 1.17 - 1.07 \times h^{-1.04} \tag{1}$$

where FV is the ventilation factor and h is the height of the opening in two opposite eaves (cm). Note: For uninsulated roofs or roofs with unventilated attics, $FV = 1$.

The solar radiation absorptivity (α) is understood as the ratio of the solar radiation incident rate absorbed by a surface. This solar radiation is concentrated in the region of the electromagnetic spectrum between a wavelength of 0.2 and 3.0 μm. Thermal delay (φ) comprises the elapsed time between a thermal variation in a medium and your manifestation in the opposite surface of a constructive component subjected to a periodical heat transmission scheme. This variable depends on the thermal capacity of constructive component and the order in which the layers are arranged (see NBR 15220-3 [5]).

In conclusion, in this section it was presented an approach with regard to the problem of housing deficit in Brazil, a projection of housing demands, the housing policy in Brazil after the year 2003, culminating with a study of promotion policy of houses by the recent Governments. In a second stage, it was presented the Brazilian energy policy, finishing with an approach of the Brazilian Bioclimatic Zones, with different climates prevalent. In this section it was still presented the Brazilian legislation related to the study of thermal performance in buildings, more precisely, the standards NBR 15220 and NBR 15575, as well as the regulation, RTQ-R, whose function is to complement the two standards with regard to the issue of energy efficiency in buildings.

In the following section it is presented a study related to the thermal comfort, their methodologies and a study of night ventilation and their contributions to the thermal comfort in Brazil and in the world.

3 Thermal Comfort and Nocturnal Ventilation

3.1 Thermal Comfort Models

The first studies of thermal comfort in buildings were recorded in the period prior to the Christian era, Vitruvius [31], in 1st century BC, which already, at the time, made reference, in your architecture considerations, of the climate in buildings projects by assigning references to healthy and solar orientation of the buildings, i.e., presented a concern with the thermal comfort in buildings.

The building thermal comfort has been object of several studies in the last decades by a significant number of researchers, as described in Table 10, as example.

The adoption of passive systems that can provide thermal comfort and reduced energy consumption in residential buildings as well as maintain the minimum indoor air quality satisfactory, it becomes critical to users of these buildings.

Peeters et al. [44] showed that the thermal comfort is the result of a combination of both environmental and human body parameters. Nicol and Humphreys [36], sustained that the fact of the internal climate be satisfactory in a building, makes it

Table 10 Previous national and international researches in thermal comfort

References	Main outputs
Houghten and Yaglou [32]	The authors presented an equation for thermal comfort and defined several thermal comfort zones
Fanger [33]	The author presented the PMV (Predicted Mean Vote) and PPD (Predicted Percentage of Dissatisfied) indices
Araújo [34]	Parameters of thermal comfort for users of school buildings
Hackenberg [35]	Comfort and heat stress in industries in addition
Nicol and Humphreys [36]	Adaptive thermal comfort and sustainability
Of Dear [37]	Improvement of Adaptive model of thermal comfort
Olessen and Parsons [38]	Confirmed the validity of the ISO 7730 only in laboratory studies
Gouvêa [39]	Thermal comfort in clothing industry
Gemelli [40]	Evaluation of thermal comfort and acoustic comfort in schools
Andreasi [41]	Evaluation of thermal comfort in hot and humid climate regions of Brazil
Lamberts and Xavier [42]	The authors gathered the, previous and current, standards of thermal comfort and heat stress, in a single document
Gagliano et al. [43]	The authors studied the variability of thermal performance of apartments in Porto, Portugal; and evaluated the influence of 4 variables in the thermal performance, of different apartments, in two periods, winter and summer. In the study, it was possible to observe the influence of occupation on the indoor air temperature. The effect was concluded on the behavior of the median, ranging between 2.6 and 2.9 °C for warmer and less warm apartments

comfortable to the point to be able to design future decisions about energy consumption and sustainability assurance.

In order to evaluate the joint effect of the variables involved in the thermal comfort, different models and methodologies are suggested by a several number of researchers. These models are developed taking into account the individual and environmental variables, such as the development of the type of activity, clothes, to then relate with the environment variables.

The evaluation of thermal comfort in this work is established according to the current standards. The existing adaptive comfort models, are specified in several standards, as ASHRAE 55 [7] and EN 15251 [8]. ISO 7730 [9] does not provide any specific model for the evaluation of the adaptive thermal comfort, however, this standard establishes a reference to the possibility of applying adaptive models. The ISO 7730 reports through experimental studies, showed that the occupants of buildings Support/accept indoor air temperatures above those laid down by index PMV of Fanger. In Portugal, an Adaptive model developed in the LNEC by Matias [45] defines, for the country, the thermal comfort in buildings. On the LNEC model, the thermal comfort index determined includes the influence of objective and subjective parameters, obtained from field studies, supplemented by sociological variables.

Among the various thermal comfort indices found, two groups of approaches worth mentioning:

- Based on heat balance (PMV and PPD);
- Adaptive approach.

3.1.1 Fanger Model (PMV and PPD)

The Fanger model is a prescriptive model of thermal comfort that includes spaces in which the occupants have metabolic rates between 1.0 and 2.0 met, and clothes with heat resistance up to 1.5 clo. This model is described by the standard EN 7730 [9] and it is based on the PMV (Predicted Mean Vote) and PPD (Predicted Percentage of Dissatisfied) indices of Fanger. Related to Fahger model, the standard provides three classes of thermal comfort (A, B, and C) with demanding levels that decrease from class A to class C. Table 11 presents the recommended values for each class, for an acceptable thermal environment for at least 65% of the occupants and taking into consideration the PMV and PPD indices. Table 11, also, presents the specifications of the occupational profile for the spaces predicted and recommended by standard.

Table 11 Comfort classes developed by Fanger [33]

Class	Space occupation profile	PPD (%)	PMV
A	People with special needs	<6	$-0.2 < PMV < 0.2$
B	New buildings or subject to rehabilitation	<10	$-0.5 < PMV < 0.5$
C	Existing buildings	<15	$-0.7 < PMV < 0.7$

Table 12 Environmental conditions using in the adaptive model

Adaptation form	Criterion adopted
Behavioural	Actions taken by the user to ensure thermal equilibrium
Physiological	Changes in heat exchange of the individual mechanisms in seeking to adjust the response of the organism to environmental conditions
Psychological	Changes of perception and reaction to sensory stimulation

3.1.2 Adaptive Approach

Then, it will be presented a brief description of the adaptive thermal comfort models listed in standards ASHRAE 55 [7], EN 15251 [8] and the thermal comfort model developed by LNEC [45].

The adaptive model considers three different forms of adaptation to the environmental conditions [46, 47]. These adaptations forms are behavioural, physiological and psychological of the environmental conditions. Table 12 shows environmental conditions using in the adaptive model for each adaptation form.

The HVAC systems offer a set of possibilities to ensure optimal conditions of comfort. The key question that arises is to obtain, maintain and control the conditions of comfort making rational use of energy and optimizing the parameters that influence the thermal comfort. These parameters can be grouped into three main categories:

- Physical parameters: In this parameters are relevant the air temperature, the average radiant temperature, air relative humidity and the relative velocity of the air;
- Organic or subjective parameters: In this parameters are included the age, sex, race, colour, or other specific features of each person;
- External parameters: This parameter includes activity levels, which are related to the metabolism of each person, the type of clothes used and the social conditions.

Among all these parameters, the parameters that most influence the thermal comfort can be described as the temperature, relative humidity, air velocity, clothes and the metabolic activity of each person.

3.1.3 Thermal Comfort Quantification

In the process of quantification of the thermal comfort, there is a relationship of interdependence between some behavioural factors, psychological and physical, whose inaccuracy implies a difficulty to scaling the thermal comfort. The general idea that the thermal comfort is based on the concept of human thermoregulation, i.e., on the human ability to maintain a constant internal temperature during the heat exchange between each person and the environment.

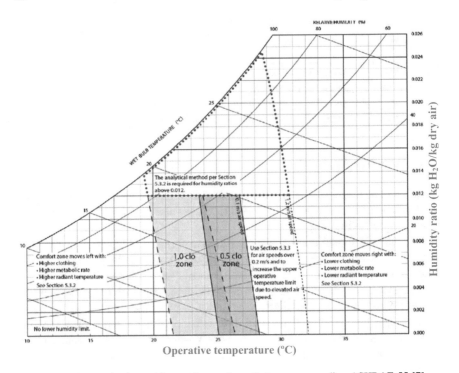

Fig. 16 Psychrometric chart with superimposed comfort zones, according ASHRAE 55 [7]

A method to quantifying the thermal comfort is presented in Fig. 16 and it was proposed by the standard ASHRAE 55 [7]. According to the standard, the environmental conditions were obtained from laboratorial experiments where the statistical analysis of the data collected allowed determine the percentage of people in light activity, i.e. in sedentary state, typical of an office building, which would be in a State of thermal comfort. Under these conditions is defined a range of metabolic activity between 1.0 and 1.3 met and thermal insulation of clothes between 0.5 and 1.0 clo, for summer and winter seasons, respectively.

The third model, the adaptive model, as happens in the prediction model of thermal comfort through the PMV index, it is also present in the same standard. This model is valid for environments without mechanical devices to cooling the air and it was provided for the situations where the user has control over the opening of windows.

3.1.4 Adaptive Model of Thermal Comfort, ASHRAE 55

The thermal comfort standard ASHRAE 55 [7] incorporated the adaptive approach as a criterion for the evaluation of thermal performance of buildings, in accordance

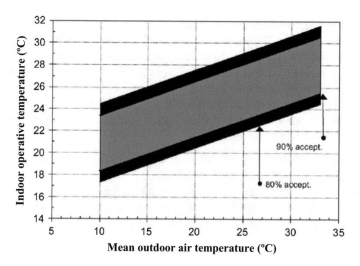

Fig. 17 Proposed adaptive comfort standard for ASHRAE 55 [48]

with the research project ASHRAE project RP-884. The adaptive model takes into account two classes of comfort corresponding to 80 and 90% of user's acceptance, as sketched in Fig. 17.

The comfort zone corresponding to 80% of acceptability, with a bandwidth of ±3.5 °C around the operative temperature of comfort, is applicable to current situations.

The comfort zone corresponding to 90% of acceptability, with a bandwidth of ±2.5 °C around the operative temperature of comfort, should be applied to buildings where it is required high levels of comfort, i.e., in the cases where the outside monthly average temperature (T_m), presents temperature fluctuations between 10 and 33.5 °C. In general, this buildings use natural ventilation as passive thermal control, which is controlled by the opening and closing of the windows by the users. These occupants presents a metabolic activity levels between 1.0 and 1.3 met and clothes to easy adapted to the indoor and outdoor environmental conditions.

The model allows comparing the average monthly temperature of the outside air with the operative temperature.

The mathematical expression represented by Eq. (2) was proposed by Brager and Dear [46] and defines the calculation of operative comfort temperature (T_{OC}) on the basis of the monthly average exterior temperature (T_m):

$$T_{OC} = 0.31 \times T_m + 17.8 \tag{2}$$

where T_{OC} is the operative temperature of comfort (°C) and T_m is the average exterior temperature (°C).

According to several authors, as Nicol and Humphreys [49] and Tuner [50], the ASHRAE standard 55 [7] no specific quantitatively the monthly average temperature present in the equation used to calculate the operative comfort temperature (T_{OC}), i.e., leaves doubts if this refers to the last 30 days of the civil calendar month or the data contained in climatological standards.

The ASHRAE standard 55 [7] admits, in simplified form, the use of indoor air temperature (T_i), as operative temperature approximation of comfort, *Toc*, if fulfilling the following conditions:

- Absence, in the buildings, of radiant panels of heating or cooling;
- Solar factor of glazing lesser than 0.48;
- Absence of a heat source in the compartment under examination;
- The heat transmission coefficient (U_m) of the glass and exterior wall openings obey the following relationship (Eq. 3):

$$U_m = \frac{50}{T_{d,i} - T_{d,s}} \tag{3}$$

where U_m is the average value between the thermal transmission coefficient of the glazed and the outer wall (W/m^2 °C), $T_{d,i}$ is the sizing indoor temperature (°C) and $T_{d,s}$ is the sizing outdoor temperature (°C).

3.1.5 Adaptive Model of Thermal Comfort, EN 15251

EN 15251 [8] standard presents a criteria to evaluate the indoor environmental quality of buildings in three categories The first category, for high level of expectation, is recommended for environments occupied by people sensitive and fragile; the second category, a normal level of expectation, should be used for new buildings and the third category, for moderate level of expectation, should be adopted for new and existing buildings.

The standard includes an adaptive method for evaluating the thermal comfort whose validation takes place in compliance with the following conditions:

- The buildings shall be free of mechanical refrigeration equipment, however, in summer season the passive cooling is possible, since the ventilation flow rates and the consumptions are relatively small;
- The temperature control should preferably be done by the opening and closing of the windows;
- The metabolic activities of the occupants must be sedentary with levels between 1.0 and 1.3 met;
- It is possible the use a mechanical heating system, provided that this system does not include a mechanical air treatment system

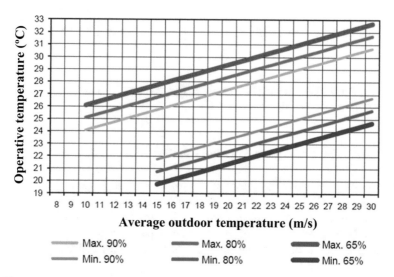

Fig. 18 Thermal comfort diagram of EN 15251 [8]

- The users are not allowed to establish "rules" of the clothes adjustment in order to better adapt to the indoor environment.

EN 15251 [8] standard establishes three comfort temperatures ranges, one for each level of expectation. The first level presents temperatures between 21.7 and 30.7 °C for an acceptability of 90%; the second level between 20.35 and 31.7 °C to an acceptability of 80%; and the third level between 19.75 and 32.7 °C to an acceptability of 65% of the users, as shown in Fig. 18.

The last version of standard EN 15251 [51] defends the adaptive model based on the experimental results of the project Smart Controls and Thermal Comfort (SCATs) presented by Nicol et al. [52] and McCartney et al. [53]. The adaptive model of standard EN 15251 applies to spaces without limitation on the number of occupants and the term adopted to calculate the operative comfort temperature (T_{OC}) is established on the basis of the exterior average weighted temperature, T_m. The expression of the EN 15251 adaptive model [8] is represented by Eq. (4):

$$T_{OC} = 0.33 \times T_m + 18.8 \tag{4}$$

where T_{OC} is the operative temperature of comfort (°C) and T_m is the average exterior temperature (°C). The outside average weighted temperature (T_m) is determined on basis of the values of daily average temperatures of the preceding week, as presented in Eq. (5). The use of the exterior average weighted temperature value (T_m), adopted the principle that the characteristic time interval for a person can fully adjust to climate change is approximately one week.

$$T_m = \frac{T_{n-1} + 0.8 \times T_{n-2} + 0.6 \times T_{n-3} + 0.5 \times T_{n-4} + 0.4 \times T_{n-5} + 0.3 \times T_{n-6} + 0.2 \times T_{n-7}}{3.8}$$

$$(5)$$

where T_m is the average weighted outdoor Temperature (°C) and T_{n-i} is the average temperature outside of the last day (°C).

3.1.6 Adaptive Portuguese Thermal Comfort Model, EN 15251

Another adaptive model available in literature was developed by the National Laboratory of Civil Engineering (LNEC) in Portugal, to the conditions for Portuguese thermal comfort in buildings. In this model, the thermal comfort index given includes the influence of environmental parameters and those relating to thermal perception, obtained on the basis of field studies and complemented by sociological factors.

The study was developed by Matias [45], supported by in situ measurements of the thermal comfort parameters and by the conducting surveys of occupants for subsequent validation of the experimental results. In this study were set two temperatures of comfort obtained as a function of exterior average weighted temperature (T_m) and calculated based on Eq. (4). The two comfort temperatures refer to 2 distinct situations:

- Buildings with HVAC systems enabled;
- Not air-conditioned buildings, for not having any type of HVAC system installed, or existing systems are disabled.

Figure 19 shows the recommended values of comfort temperature in function of the outdoor temperature, using in the LNEC model. According the work developed by Almeida [54], the research presented by Matias [45] includes a comfort zone for 90% accessibility, defined with a bandwidth of ±3 °C around the comfort temperature (see Fig. 19). Figure 19 shows the air-conditioned spaces and the limits temperatures in winter and summer, respectively, 15 and 31 °C. Clearly it is possible to observe a greater tolerance to extreme limits of occupant thermal comfort in non-air conditioned buildings, comparing with the air-conditioned buildings. According to Matias [48], the indicated temperatures limit should only be considered in interior environments where exists a high possibility of personal or environmental adaptation, only available to residential buildings.

In the experimental study developed by Matias [45] in service and residential buildings, in order to assess the conditions of thermal comfort, several environmental parameters were measured (indoor and outdoor), such as the indoor air temperature (T_i) and the average radiant temperature (T_{mr}). From these temperatures, it was possible to determine the operative comfort temperature (T_{OC}). In Fig. 20 it is presented the correlation obtained between the average values of the indoor air temperature (T_i) and the operative comfort temperature (T_{OC}), in the

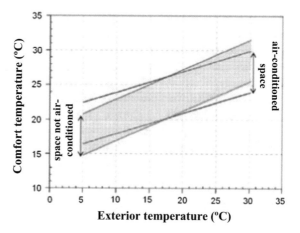

Fig. 19 Recommended values of comfort temperature in function of the outdoor temperature. Adapted from [34]

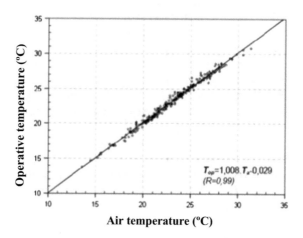

Fig. 20 Correlation between the mean values of the operative temperature and the air temperature. Adapted from Araújo [34]

in-field studies developed [45]. The experimental results reveal a dependent relationship between the two parameters correlated (with a linear coefficient of Pearson equal to $R = 0.99$), and confirm that, on current situations, it is possible to use the indoor air temperature as a good approximation of the operative temperature of comfort (T_{OC}).

Finally, it is important to be in mind that the LNEC model [48] developed by Matias [45] for the evaluation of thermal comfort conditions were obtained on service buildings, such as, offices, schools and nursing homes, not extending to the residential sector, only for 4 multi-family housing.

The evaluation of the buildings built with gypsum blocks, in Brazil, with the aimed to analyses the thermal comfort is main goal to achieve, and the model that will be adopted is the adaptive method described by ASHRAE 55 standards and ISO 7730. The gypsum house under study not presents any mechanical cooling system and the exterior monthly average temperature varies between 10 and 33.5 °C;

the occupants have activity level between 1.0 and 1.3 met and can adapt your clothes to the interior and exterior environmental conditions. Emphasis will be given to the definition of the users profile, location and conditions in service of the prototype testing. The operative temperature of comfort will be calculated by Eq. (2).

The adaptive model presented in ASHRAE 55 [7] was used, recently, in a work developed, in Portugal, by Curado [55] for the calculation of thermal comfort, in winter and summer seasons, of standard apartments in the city of Porto, and having considered a comfort class of 80% of the occupants.

3.2 Night Ventilation (Natural Ventilation)

Natural ventilation is the process of supplying air to and removing air from an indoor space without using mechanical systems. It refers to the flow of external air to an indoor space as a result of pressure differences arising from natural forces.

Gratia et al. [56] shows that, in the majority of cases, natural ventilation can be enough to ensure thermal comfort in buildings occupied, with some effort to reduce internal heat generation through a correct personal management and equipment choice. On the other hand, a study developed by Santamouris [57] presents the advantages associated to the use of natural ventilation on indoor thermal conditions, providing these benefits to a number of than 3 billion of people worldwide, especially those entered in the middle and lower classes. According to Schiffer and Fleet [58] the countries with hot and humid weather, for example, Brazil, the natural ventilation presents itself as a passive, low-cost alternative to the periods in which the heat discomfort is evident. Ventilation provides a feeling of comfort to the occupants through the control of indoor air, respecting limits of temperature, air velocity and air humidity.

In accordance with Mazon et al. [59], the strategy associated to correct natural ventilation it is in the adjustment of the building internal climate by a controlled air exchange through the holes. The mechanism of air movement in and out of a building, according to Hunziker [60], under the action of natural atmospheric forces, is relevant to the study of the thermal comfort of the users. One of the benefits of night ventilation is the promotion of building cooling by removing the thermal load absorbed by buildings due to the exposure of the building the solar radiation, as well as the thermal gains produced inside the buildings by the presence of users, electrical equipment's, artificial lighting, etc. [61]. In these cases, high rates of ventilation tend to provide internal temperatures very close from the outside, removing part of the existing thermal load in indoor environments [62, 63].

Gratia et al. [56] and Andreasi [64] show that the maximum acceptable air velocity on indoor environment is 0.8 m/s, and this air velocity causes a cooling sensation of approx. 3 °C. These cooling sensations through the vents for air velocity up to 0.1 m/s may only be felt at temperatures below 18 °C, which is undesirable. In the case of higher temperatures, for air velocities above 0.2 m/s,

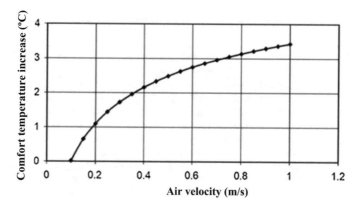

Fig. 21 Influence of air velocity on comfort temperature. Adapted from Nicol [65]

the users can take advantage of the cooling sensation through the speed of air passing through the body. A research developed by Nicol [65] shows the benefits that natural ventilation could provide, in tropical climates with temperatures above 30 °C, provided that the local that presents this temperatures had an increase of mechanical ventilation systems.

Among the factors that allow this tolerance limit of temperature variation on comfort zone, are the processes of adaptation of the users to the environment and their possibility to look for the place with the best ventilation designed by the equipment. This adaptability may increase the comfort temperature in 2 or 3 °C. Figure 21 shows the an increase in the air velocity of 1 m/s is capable of raising the temperature of comfort to values higher than 3 °C. Therefore, it is considered essential the use of natural ventilation to improve thermal sensation of the building users, even in cases that the outside temperature is higher than 29 °C.

3.2.1 International Contributions to Nocturnal Ventilation

Several international studies have justified the efficiency associated to the use of nocturnal ventilation with low cost to users. Some of these studies are presented summarily in Table 13.

This analysis was the base of the study presented in [68, 69]. In this work was presented a study of thermal comfort and night ventilation of a house built with plaster blocks in the city of Petrolina. The economic situation of a good portion of the lower middle class in Brazil, without habits to cool in the hottest stations, assumed the importance of adopting the Adaptive methodology in approaching the study of thermal comfort.

The following section makes an approach importance of plaster production in Brazil and its applicability to building construction, as well as displays succinctly a scenario of international standards regarding this product. It is also described some

Table 13 Major international contributions on use of nocturnal ventilation

Author	Major contributions
Givoni [26]	Nocturnal ventilation maintain the interior temperature below 24.5 °C, although the temperature outside be 38 °C
Artman et al. [51]	The study proved the efficiency of nocturnal ventilation, in summer, in countries of Northern Europe and the British Isles. This efficiency, however, decreases in Central and East Europe, namely in Portugal, Spain, Greece and Turkey, while maintaining an acceptable level. Register as an example, the Lisbon city, with 200 nights a year when one considers the potential of nocturnal ventilation cooling effective. In that city, in just 15 nights of the year the night ventilation cooling potential, not efficient. In the same study, the North of Portugal is appointed as a region with a strong potential of night ventilation, compared to the level of the Centre of Europe
Shaviv et al. [66]	In this study conducted in 4 cities of Israel it was evaluated the influence of thermal inertia and nocturnal ventilation in indoor air, maximum temperature in summer, of a housing building. In this study it was considered 4 inertia thermal classes (light, medium, semi-heavy and heavy) and 4 levels of nocturnal ventilation (without night ventilation, natural ventilation with hourly rate of 5 h^{-1} and forced ventilation with rates of 20 and 30 Rph). The results show that in a building with optimized thermal inertia and with air renewal rates suitable for night ventilation it is possible to obtain a reduction between 3 and 6 °C in indoor air temperature values
Santamouris et al. [67]	The study assesses the impact of ventilation on energy requirements in buildings depending on the variation of ventilation rates and features of buildings. In this research, it was monitored 214 housing buildings in Greece for evaluation of the impact of night ventilation on cooling energy needs. The results obtained show the impact of night ventilation depends of the air renewal rate adopted and constructive characteristics of the monitored buildings. In quantitative terms, the night ventilation presents a maximum reduction of the cooling annual requirements of approx. 40 kWh/m^2, with the average annual reduction equal to 12 kWh/m^2

Adapted from Curado [55]

aspects of sustainability of gypsum constructions, although this is not the matter that constitutes the main objective of this work.

4   Gypsum Importance in Brazilian Building Construction

4.1   Examples of Gypsum Houses

According Pires Sobrinho [70], the vertical building construction with plaster blocks should not be considered technological innovation in Brazil, since there are records of cases buildings, with over 12 floors, and with more than 10 years in Jaboatão dos Guararapes-PE. It should be noted that vertical building construction with plaster blocks is limited to the Brazilian states of the Northeast region, more specifically in Pernambuco, Ceará and Sergipe.

The use of gypsum plaster in building blocks, in Brazil, is still a restricted item, due to lack of knowledge on technology. Another factor that restricts the use of the blocks is the location of the deposits of gypsum, a material that gives rise to the plaster, which are located in the West of the State of Pernambuco, located far from the large consuming centers, potentially located in the southeast of the country. However, the interest in your study is determined by the annual consumption growth that reaches about 8% per year in Brazil.

Although a growing demand in the use of plaster blocks in the internal seals in multiple buildings, especially in big cities of the Northeast and even inside, there is furthermore a great unknown regarding your production technology, as well as your behaviour, caused by the lack of national standards, considering fundamental research carried out in Brazil on the subject [71]. Figures 22, 23, 24 and 25 show some examples of building constructions that employ gypsum blocks in different Brazilian states, such as Pernambuco, Petrolina, Ceará and Rio de Janeiro.

Plaster is a binder, which compared with other materials as whitewash and Portland cement, can be considered much less aggressive to the environment. While, in the manufacture process, the Portland cement emits CO_2, the plaster launches water molecules in the atmosphere. Production of Portland cement requires high temperatures, while gypsum plaster can be prepared with temperatures of about 150 °C. Although these ambient advantages, gypsum plaster is not very often applied in Brazilian buildings and an effort has been made to increase the consumption of this material in housing. Gypsum, abundant in nature and found in several deposits in northeast Brazil (plaster pole of Araripe, Pernambuco), can be used for manufacturing masonry units and renderings, resulting in a great added value for the local economy and environment.

The Araripe region is located in the state of Pernambuco, about 700 km from the capital Recife and in the same radius of the main capitals of the Brazilian Northeast. Covering a total area of 7074.60 Km^2 (see Fig. 26), with 235,446 inhabitants.

Fig. 22 Examples of building constructions that employ gypsum (flats built in gypsum plaster, in Petrolina and Belem do São Francisco-PE)

Fig. 23 Examples of
building constructions that
employ gypsum (gypsum
house under construction,
Gravatá-PE)

Fig. 24 Examples of
building constructions that
employ gypsum (building
with inner walls in gypsum
plaster, São Paulo)

The Araripe region includes the cities of Araripina, Bodocó, Ipubi, Ouricuri and Trinity [72]. This region is responsible for approximately 90% of the national gypsum production and 81% of the gypsum plaster production [73].

The lands of the region of Araripe, according to Luz and Lins [75], are considered one of the best lands related to the ore quality in the world and feature excellent mining conditions. According to the National Department of Mineral production-DNPM/PE, the industrial pole of Pernambuco is responsible for more than 85% of all gypsum Brazilian production [76].

In Brazil, as described Mancino [77], there are large reserves of high purity gypsum with 95% of these deposits concentrated Northwest of Pernambuco state and that part of this plaster is used for orthopaedic and dental industry.

Fig. 25 Examples of building constructions that employ gypsum (building with internal and external walls in gypsum plaster, Rio de Janeiro)

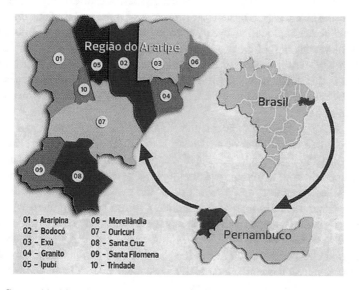

Fig. 26 Geographical location of Araripe region [74]

4.2 Importance of Gypsum Construction in Brazilian North-Eastern

The Brazilian north-eastern region, according João Pinheiro Foundation [3], presents a high relative percentage (91.3%) of housing deficit. Building construction applying gypsum plaster, a raw material abundant in Brazil, appears as a relevant factor with enough potential to mitigate this habitational deficit and become another

Table 14 Gypsum plaster consumption "Per capita" (2005) [74]

Countries	Consumption "per capita" (kg)	Global position
Canada	315.6	1
Spain	189.3	2
Iran	145.9	3
Thailand	82.1	4
United States	81.1	5
France	76.4	6
Mexico	71.8	7
Japan	43.4	8
Brazil	13.0	9
China	7.0	10

important alternative available to the Brazilian market. According to the Union of Pernambuco Industries of Gypsum Plaster-SINDUSGESSO [74], Pernambuco, situated in the northeast of Brazil, has a reserve of approximately 1.2 billion tons with potential for exploitation by a period estimated of 500 years. Brazil despite having such reservation of the product, according to the Union, is located in the 9th position in the world rankings, having embryonic consumption countries highlighted as major consumers such as Canada, Spain and Iran, whose consumption "*per capita*" are on the order of 315; 190 and 145 kg/(person.year), respectively (see Table 14).

In Brazil, the building construction process with gypsum blocks is in an embryonic phase and orphan of a relevant national legislation on the subject [27]. According to Pires Sobrinho et al. [78], some initial benefits are obtained in a building construction involving gypsum blocks. The authors found in their research a reduction in labour cost and a reduction of the overload and time of execution.

Recent experimental results with Brazilian gypsum blocks, presented by Santos [79], in an intensive experimental campaign developed in the Laboratory of Building Physics (LFC) of FEUP, in Portugal, showed that the gypsum blocks present good constructive ability from the point of view of thermal behaviour, with low coefficients of thermal conductivity ($\lambda = 0.28$ W/m °C), important to their employment in countries of hot weather.

4.3   Sustainability of Gypsum Constructions

Results of the study presented by Pires Sobrinho et al. [78] demonstrate some benefits obtained in buildings built internally with plaster masonry when compared to conventional brick masonry (see Table 15). In this study it was presented a comparison between the cost, by volume, of the armour concrete applied in different typology of buildings and the load on the building foundation. The typology

Table 15 Comparative financial and percentage of plaster masonry and brick masonry [78]

Type	Armor in the structures			Concrete in the foundation			Load on the foundations	
	(Ton)	R$	%	(m³)	R$	%	(Ton)	%
6	17	110,500		41.9	50,280		1386	
6 g	15	97,500	−11.8	30.9	37,080	−35.6	1158	−16.4
14	55	357,500		166.02	199,224		3927	
14 g	49	318,500	−12.2	125.79	150,948	−32.0	3315	−15.6
22	114	741,000		322.96	387,552		6282	
22 g	101	656,500	−11.4	244.62	293,544	−32.0	5309	−15.5

adopted, 6 (brick wall); 6 g (dry wall); 14 (brick walls); 14 g (drywall); 22 (brick walls) and 22 g (Sheetrock) refers to buildings of 6, 14 and 22 floors. According to the authors, the cost analysis was performed based on the values used as reference in official public costs of urban cleaning company of Recife (EMLURB Jan/2009), organ of the city of Recife, in the state of Pernambuco.

It is possible to observe in the analysis of Table 15, a reduction in financial cost of structural armour, in percentages, between 11.4 and 12.2%; for the volume of concrete foundations a reduction between 32.0 and 35.6%, and finally, a reduction of load on the foundations between 15.5 and 16.4%, factors of extreme relevance to civil engineering.

Related to energy issue, a study developed by Abeysundara et al. [80] showed that the embodied energy is one way of measuring the environmental impact of buildings, an indicator of the sustainability of buildings. This method of calculation accounts the entire energy used in manufacturing, transport and use of the material, in the construction.

Table 16 presents a study developed by Tavares [81], which show a comparative analysis of embodied energy estimate taking into account the superstructure armour, concrete foundation and internal masonry in apartments of buildings scaled with 6, 14 and 22 floors, with plaster walls (g) and ceramic brick, based in unit values of embodied energy.

The buildings using plaster blocks presents, in accordance with Table 16, an energy reduction rates between 16.6 and 17.3% in relation to the traditional constructive mode with ceramic blocks, reaching greater benefits for up to six floors, with regard to the structure of the building.

4.4 Durability Evaluation of Plaster Buildings in the Northeast Brazil

Although the study of durability of buildings built with gypsum blocks is not the main objective of this work, it is presented a brief description of a building/

Table 16 Comparative data in function of the estimate energy embodied [81]

Type	Superstructure armour			Concrete foundations			Internal masonry			Total	
	Weight (ton)	Energy embodied (MJ)	%	Vol (m³)	Energy embodied (MJ)	%	Weight (ton)	Energy embodied (MJ)	%	Energy embodied (MJ)	%
6	17	510,000		41.9	311,736		434.88	1100		822,836	
6 g	15	450,000	−11.8	30.9	229,896	−26.3	300.24	975	−11.4	680,871	−17.3
14	55	1,650,000		166.0	1,235,189		1014.72	2567		2,887,756	
14 g	49	1,470,000	−10.9	125.8	935,878	−24.4	700.56	2275	−11.4	2,408,153	−16.6
22	114	3,420,000		323.0	2,402,822		1594.56	4034		5,826,857	
22 g	101	3,030,000	−11.4	244.6	1,819,973	−24.3	1100.88	3575	−11.4	4,853,548	−16.7

Fig. 27 Prototype of a gypsum house with 8 years, in ITEP. *Source* ITEP (2015)

prototype built in the last years and subject to natural weather conditions. Should be mentioned that this type of study does not involve changes of hygrothermal behaviour that could compromise the development of this work.

Figure 27 shows a prototype built with blocks of gypsum plaster for a study developed at ITEP-Technological Institute of the State of Pernambuco with eight years of construction. According to the reports presented by technical researches, the pathologies observed in the prototype, 8 years after its construction and subject to natural weather conditions, are only some detachments of plaster glued on windows demarcation and painting detachments on the lower surface, at a height of 50 cm from the ground, resulted of rain water spatter.

4.5 Brief Literature Review of Gypsum Application

In use for more than 5000 years, gypsum is still reserved a place among indispensable building materials on account of its outstanding properties. Gypsum material is a typical eco-type and energy-saving material. Natural gypsum has long been used as raw materials, which consumed a great deal of natural resources. Gypsum is one of the most ubiquitous compounds found both in nature as well as on buildings.

Gypsum, sodium sulphate, Na_2SO_4, can dehydrate to a lower hydrate, the hemihydrate, and to an anhydrous phase, anhydrite. Calcium sulphate, $CaSO_4$, can appear as three distinct minerals: gypsum (dihydrate), bassanite (hemihydrate) and anhydrite (anhydrous). These can be summarized as follows: there are two stable phases, gypsum and anhydrite, and two metastable ones, hemihydrate and the so-called soluble anhydrite.

Among the positive qualities of gypsum valued in the construction business are its good thermal and sound insulation, light-weight, ease of application, resistance to fire, hydrometric dimensional stability, smooth surface, good compressive strength, and neutral pH. Further, it is a readily available natural resource. The main defect of gypsum is its lack of resistance to moisture, making it unsuitable for use on exposed exteriors. For this reason, much research on gypsum has been aimed at enhancing its resistance to atmospheric conditions.

Moisture is a major cause of deterioration in buildings and its materials, in particular, the water content and relative humidity variation in walls, where the lack of care in the assessment of hygrothermal behaviour of buildings can lead to occurrence of pathologies.

Buildings hygrothermal performance has been a subject of great demand, by construction professionals, as regards the new construction and rehabilitation. The proper choice of materials, building elements and building technology is increasingly supported by numerical simulation that replicates the buildings' behaviour. In that case, the materials hygrothermal properties represent the fundamental basis for achieving these hygrothermal simulations.

To achieve the ideal conditions of hygrothermal comfort inside a building, it is necessary to have greater control of the factors that influence those conditions, both interior temperature and relative humidity, which are affected by climatic conditions and the use of the buildings, as well as the characteristics relating to the geometry, location and façade composition.

Sayil and Çolak [82] analysed the mechanical strength and moisture resistance of hemi-hydrate gypsum ($CaSO_4 \cdot \frac{1}{2}H_2O$) combined separately with acrylic, epoxy and silicone resins. The results obtained showed that the best performance among the different resins was displayed by the specimens containing silicone resin. The moisture durability of gypsum compounded with silicone resin was less than 100%, who demonstrated that gypsum modified in this manner can be utilized as an outdoor material.

Okino et al. [83, 84] studied the incorporation of cement to particleboard and they obtained better physical and mechanical properties of cement particleboards than those reported by Frick [85] for woodgypsum composite boards. The results obtained by Okino et al. [83, 84] presented lower water absorption, better dimensional stability and higher mechanical strength.

Deng et al. [86] studied the influence of an organosilicon agent on the physical and mechanical properties of the boards. Some years later the authors (Deng and Furuno [87, 88]) added polypropylene fibers, jute fibers and coconut fibers to gypsum particleboard. However, the results obtained by these authors were not conclusive, regarding the water resistance of gypsum particleboard.

Feng-Ging et al. [89] analysed the water resistant gypsum block from flue gas desulfurization gypsum. For this purpose they used a water resistant agent prepared from granulated blast-furnace slag, high calcium fly ash and some additives, which was mixed with calcined desulfurization gypsum to form modified gypsum powder which is used to produce water resistant block. The results obtained showed that a significantly improve of the gypsum products to the water resistance.

Bülichen and Plank [90] analysed the water retention effectiveness of a commercial methyl hydroxypropyl cellulose sample in a gypsum plaster. The results showed that the water retention capacity decreased significantly in the calcium sulphate system. Thus, higher dosages of methyl hydroxypropyl cellulose were required in the gypsum plaster to attain water retention values comparable to those in cement.

The use of gypsum plaster in building blocks, in Brazil, is still a restricted item, due to lack of knowledge on technology. Another factor that restricts the use of the gypsum blocks is the location of the deposits of gypsum, a material that gives rise to the plaster, which are located in the West of the State of Pernambuco, located far from the large consuming centers, potentially located in the southeast of the country. However, the interest in your study is determined by the annual consumption growth that reaches about 8% per year in Brazil (Peres et al. [91]; Baltar et al. [92] and Araújo [93]).

Currently, gypsum based blocks are destined to residential and commercial buildings interior walls, reinforced concrete structures or similar constructions (see Fig. 28). In the Northeast of Brazil, the construction industry has also applied gypsum based blocks in façades, despite the lack of legal framework and regulation (Pinheiro [94]).

Nolhier [95] showed that the idea of making prefabricated elements of seal with plaster-based materials, constituting a wall, came up as the first use of plaster. Archaeological excavations have revealed the use of this type of prefabricated in Syria, in the 6th millennium BC, and in Paris in the 3rd century AC.

According to Mancino [77], the Egyptians, about 5000 years ago, made plaster outdoor burning, turning it into a powder. Subsequently, they mixed this powder with water to make a mineral paste whose goal was to link building blocks. These blocks were used to build the enormous Egyptian monuments, many of which survive to this day, as the magnificent *Sphinx* and the tomb of *Tutankhamun*.

The manufacture of plaster was rudimentary and empirical, however, *Lavoisier* in 1768, presented at the French Academy of Sciences the first scientific study on the phenomena that constituted the basis for plaster preparation.

According to the author, in Iran, the most common application is in plaster mortar, however plaster blocks are also popular and used to construct internal partitions in residential buildings. In Europe, the use of plaster in construction has become quite popular in the 18th century. At that time, in Paris, the use of plaster in construction was quite widespread. In France, 75% of the hotels and the public buildings use wood panels and gypsum mortar, and 95% of the new buildings and building rehabilitations were made with gypsum plaster. In Spain, according to Atedy [96], the use of plaster came through the Arabs, being evident in Muslim and Moorish arts, especially in Aragon, Toledo and Seville. Therefore, it can be affirmed that the plaster is one of the oldest building materials on the planet. United Arab Emirates also manufactures plaster blocks for use in construction and there are already companies that produce them. A manufacturing unit of plaster blocks, created in 1978 in Abu Dhabi, currently produces 100,000 m^2/day (see Mancino [77]).

Fig. 28 **a** Standard gypsum plaster blocks; **b** hydro gypsum plaster blocks; **c** gypsum plaster blocks reinforced with fiberglass; **d** gypsum plaster blocks reinforced with fiberglass and waterproofing

Countries such as Spain, France and United States use in large scale construction technologies with the cast, particularly in the execution of internal seals, linings and coatings. Is assigned to France the origin and development of the technology of vertical seals with gypsum blocks [97].

Today, French plaster industries represent about 95% of the country's production of all products in the construction. In 2005, consumption of gypsum board in France was more than 260 million of m^2, approximately, 4 m^2 per inhabitant.

According to the Brazilian Company Gypsum [98], there are market studies to carry out residential projects built in plaster for African countries, including Mozambique and Angola. A demonstration of four dwellings in these countries led to a sale of 600 homes. In Managua, capital of Nicaragua, another country of warm weather and in development, a pilot project was implemented with the construction of 20,000 houses with 40 and 60 m^2 made entirely in gypsum plaster.

4.6 International Standardization on Plaster

According to AENOR, the Ministry of development, within the framework of its powers, has created patterns of Construction technology (NTE), which includes, among others, the regulations on the quality of the design, construction, control, evaluation and maintenance of the work units involved in gypsum products, such as those relating to interior seals and liners (see Table 17).

In resume, this section allowed an exploratory study about the constructions with gypsum blocks in Brazil and in the world. It was possible to present a relative breakthrough of this kind of construction, mainly in the north-eastern region of Brazil, especially in the states of Pernambuco, Ceará and Sergipe. Despite being a constructive material of great importance from the point of view of sustainability, as well as the energy issue, Brazil still behaves as an embryonic consumer in relation to the consumption records of other countries like Canada and Spain.

Table 17 International standards for plaster and gypsum blocks

Standard	Year	Origin	Standard guideline
UNE EN 12859 [99]	2008	Spain	Gypsum blocks—definitions, requirements and test methods
UNE EN 12860 [100]	2001	Spain	Gypsum based adhesives for gypsum blocks—definitions, requirements and test methods
UNE EN 15318 [101]	2007	Spain	Design and application of gypsum blocks
UNE EN 13279-1 [102]	2008	Spain	Gypsum binders and gypsum plasters
P72-202 [103]	1994	France	Verticaux de plâtrerie Ouvrages ne nécessitant pas l ' application d'un enduit au plâtre-Exécution des cloisons en carreaux de plâtre

The durability issue, addressed residually in this section constitutes a fundamental aspect for the study and applicability of gypsum blocks in large scale, not being explored in this work, not interfere with the proposed objectives, getting as a topic to be explored in future studies.

5 Conclusions

The Brazilian north-eastern region presents a high percentage of housing deficits. Building construction applying gypsum plaster, a raw material abundant in Brazil, appears as an interesting alternative with enough potential to mitigate the habitational deficit.

In this work it was presented an analysis the housing deficit in Brazil, who indicates a clear need for new housing construction in Brazil. It was discussed the policy of house promotion by the Brazilian Government in the last years, presents an exhibition of energy policy in Brazil and finishing with an approach of the Brazilian bioclimatic zones with very different prevalent climates.

Also, it was presented a brief analyses of the thermal comfort, comfort-related concepts and their evaluation parameters, presents the models proposed by the standards comfort ASHRAE 55, EN 15251 and ISO 7730, deals with the LNEC model and its similarities with ASHRAE 55 model and culminates with the presentation and justification of the model adopted to delineate the comfort in this research works. Finally, a study on night ventilation and their contributions to the comfort as well as a synthesis of the studies related to the topic was discussed.

The hygrothermal analysis, reflected on the thermal comfort and night time ventilation in Brazilian gypsum plaster houses, was based on the adaptive model described in ASHRAE 55 and ISO 7730 standards for the evaluation of thermal comfort

Finally, it was presented an approach related to the plaster; make an exploratory study about advancing this constructive modality, particularly in the north-eastern region of Brazil and their projections to the country. Some questions addressed to the durability of gypsum houses has not been analysed in this work and the authors suggest their study in future researches.

References

1. Census 2010.: http://www.ibge.gov.br/. Access in Jan 2017
2. Mineral Contents.: DNPM/PE (2006)
3. João Pinheiro Foundation.: Statistical and information center. Housing deficit in Brazil, selected geographical micro-regions and municipalities. Agreement UNDP/Ministry of cities, Belo Horizonte (2005)

4. Cedeplar/UFMG-Regional Development and Planning Center.: Study to the Ministry of cities Project-demand demographic projection housing the housing deficit and subnormal settlements (2007)
5. NBR-15220-3.: Thermal performance of buildings—Part 3: Brazilian Bioclimatic zoning and construction guidelines for detached houses of social interest. Brazilian Association of Technical Standards. Rio de Janeiro (2005)
6. NBR 15575-5.: "Residential buildings—performance". Requirements for roofing systems. Brazilian Association of technical standards (2013)
7. ASHRAE-55.: Thermal Environmental Conditions for Human Occupancy. American Society of Heating, Refrigerating and Air-conditioning Engineers, Inc., Atlanta (2010)
8. EN 15251.: Indoor environmental input parameters for design and assessment of energy performance of buildings-addressing indoor air quality, thermal environment, lighting and acoustics. CEN, Brussels, Belgium (2007)
9. International Organization for Standardization: ISO 7730, Ambiances thermiques modérées – Détermination des indices VMP PPD spécification et des conditions et de confort thermique. International Standardisation Organisation, Geneva, Suisse (2005)
10. Maricato, E.: Brazil, Cities: Alternatives to the Urban Crisis. Petrópolis, Vozes (2001)
11. Ministry of cities/SNH.: National housing policy. The cities Book 4, Brasília (2006)
12. João Pinheiro Foundation.: Statistical and information center. Housing deficit in Brazil 2000. Agreement UNDP/Special Secretariat for urban development (Presidency of the Republic), Belo Horizonte (2001)
13. F. Garcia and A.M. Castle, "Housing deficit grows despite the expansion of credit". Construction Situation, year 4 (Mar.), n. 1, pp. 8–11, March 2006
14. Denaldi, A.: State Housing Policies And Favelas in Brazil. Leopoldianum Magazine, Santos (2004)
15. Sustainable Brazil.: Potential Housing Market. Ernst & Young, São Paulo (2008)
16. IBGE, National Research for Sample of Domiciles-2005 Micro-Data. Rio de Janeiro, CD-ROM (2006)
17. Cedeplar/UFMG-Regional Development and Planning Center.: Study to the Ministry of cities Project-demand demographic projection housing the housing deficit and subnormal settlements (2007)
18. Law No. 11,977, of 7 July 2009. Plateau, 2014. Available at http://www.planalto.gov.br. Accessed Jan 2017
19. FGV.: Fundação Getúlio Vargas: A construction for a sustainable development—The importance of construction in the economic and social life of Brazil, FGV Projects (2007)
20. BEN-2010 National energy balance: base year 2009/energy research company. EPE, Rio de Janeiro (2010)
21. Carrières, R. R.: Thermal performance and energy consumption of office buildings in San Carlos. MSc Thesis in Civil Engineering, Department of Civil Engineering, University of Campinas, Campinas (2007)
22. Brazil Presidency of the Republic. Civil House. Cabinet Subcommittee for Legal Affairs. Decree n° 4,059 of 19 December 2001. Regulates law No. 10,295, of 17 October 2001, which provides for the National Policy for the conservation and rational use of energy, and other matters. In: Official Journal of the European Union. Brasília, DF (2001)
23. Huberman, B., Pearlmutter, D.: The life-cycle energy analysis of building materials in the Negev desert. Energy Build. **40**, 837–848 (2008)
24. Kuznik, F., David, D., Johann, K., Roux, J.J.: The review on phase change materials integrated in building walls. Renew. Sustain. Energy Rev. **15**(1), 379–391 (2011)
25. Achão, C.C.L.: Analysis of the structure of energy consumption by the Brazilian Residences [Rio de Janeiro], 103 p. MSc. Thesis-Federal University of Rio de Janeiro, COPPE (2003)
26. Givoni, B.: Effectiveness of mass and night ventilation in lowering the indoor daytime temperatures. Part I: experimental periods, School of Arts and Achitecture, Los Angeles, United States America (1997)

27. Koenigsberger, O.H., Ingerssol, T.G., Maythew, A., Szokolay, S.V.: Manual of Tropical Housing and Building. Part I: Climatic Design. Longman, London (1974)
28. Roriz, M.: Correções nas Irradiâncias e Iluminâncias dos arquivos EPW da Base ANTAC. Grupo de Trabalho sobre Conforto e Eficiência Energética de Edificações. ANTAC – Associação Nacional de Tecnologia do Ambiente Construído. São Carlos – SP (2012). http:// www.labeee.ufsc.br/downloads/. Accessed in 2017
29. NBR 15575-4.: "Residential buildings—performance". Requirements for the systems of internal and external vertical seals. Rio de Janeiro, Brazilian Association of technical standards (2013)
30. RTQ-R, Ministry of Industry and Foreign Trade.: Quality technical regulation to the level of energy efficiency of Residential Buildings. Eletrobrás (2010)
31. Treaty of architecture, Vitruvius
32. Hougthen, F.C.. Yaglou, C.P.: Determining lines to equal comfort, and determination of the comfort zone. ASHRAE Trans. 29 (1923)
33. Fanger, P.O., Toftum, J.: Extension of the PMV model to Non-air-conditioned buildings in warm climates. Energy Build. (2006)
34. Araújo, V.M.D.: Parameters of thermal comfort for users of school buildings in the Northeastern Brazilian coast. Doctoral thesis, Faculty of Architecture and Urbanism of the University of São Paulo, São Paulo (1996)
35. Hackenberg, A.M.: Comfort and heat stress in industries: research carried out in the regions of Joinville, SC and Campinas, SP. 270 f. Doctoral thesis in Mechanical Engineering, State University of Campinas, SP (2000)
36. Nicol, J.F., Humphreys, M.A.: Adaptive thermal comfort and thermal standards for sustainable buildings. In: Moving Thermal Comfort Standards into the 21 Century, 2001, Windsor, UK. Proceedings. Oxford Brookes University, Oxford, April 2001
37. Dear, R.J.: The adaptive model of thermal comfort: Macquarie University's ASHRAE RP-884 Project (2004)
38. Olesen, B.W., Parsons, K.C.: Introduction to thermal comfort standards and to the proposed new version of EN ISO 7730. Energy Build. 34(6), 537–548 (2002)
39. Gouvêa, T.C.: Evaluation of thermal comfort: An experience in manufacturing industry, 164 f. Dissertation (Master in Civil Engineering). Department of Civil Engineering, University of Campinas Faculty of Civil Engineering, architecture and urbanism, Campinas (2004)
40. Gemelli, C.S.B.: Evaluation of thermal comfort, acoustic and luminance of building sustainable and bioclimatic strategies school: the case of elementary Municipal school Frei Pacific. 2009.17 f. Dissertation (Master in Civil Engineering), Graduate Program in Civil Engineering, Federal University of Rio Grande do Sul. Porto Alegre, Brazil (2009)
41. Andreasi, A.A.: Method for evaluation of thermal comfort in hot and humid climate region in Brazil. PhD Thesis in Civil Engineering, Federal University of Santa Catarina, Santa Catarina (2009)
42. Lamberts, R., Xavier, A.A.P.: Thermal comfort and heat stress. Laboratory of energy efficiency in buildings. Florianópolis, SC (2011)
43. Gagliano, A., Nocera, F., Patania, F., Capizzi, G.A.: Case study of energy retrofit in social housing units. In: Proceedings of the Mediterranean Green Energy Forum 2013, MGEF-13, Fes, Morocco, 16–20 June 2013
44. Peeters, L., Dear, R., Hansen, J.D., Haeseleer, W.: Thermal comfort in residential buildings: Comfort values and scales for building energy simulation. Appl. Energy 86 (2009)
45. Matias, L.: Development of a model for defining Adaptive thermal comfort conditions in Portugal. Collection Theses and Research programs LNEC, ICC 65. ISBN 978-972-49-2207-2. LNEC, Lisbon (2010)
46. Brager, G., Dear, R.: Thermal adaptation in the built environment: a literature review. Energy Build. 27, 83–96 (1998)
47. Dear, R., Brager, G., Cooper, D.: Developing an adaptive model of thermal comfort and preference. ASHRAE RP-884-Final Report, American Society of Heating, Refrigerating and Air-Conditioning Engineers, Atlanta, USA (1997)

48. Artmann, N., Heinrich, M., Heiselberg, P.: Potential for passive cooling of buildings by night-time ventilation in present and future climates in Europe. In: 23rd International PLEA Conference, Geneva, Switzerland (2006)
49. Nicol, F., Humphreys, M.: Derivation of the adaptive equations for thermal comfort in free-running buildings in European standard EN 15251. Build. Environ. **45**, 11–17 (2010)
50. Turner, S.: ASHRAE`s thermal comfort standard in America: Future steps away from energy intensive design. In: Proceedings of Conference: Air Conditioning and the Low Carbon Cooling Challenge, Windsor, UK, 27–29 July 2008
51. Artmann, N., Heinrich, M., Heiselberg, P.: Potential for passive cooling of buildings by night-time ventilation in present and future climates in Europe. In: 23rd International PLEA Conference, Geneva, Switzerland (2006)
52. Nicol, F., MCCartney, K.: Smart controls and thermal comfort (SCATs) project. Final Report, Joule III Project (Contract JOE3-CT97-0066) (2000)
53. MCCartney, K., Nicol, F.: Developing an adaptive control algorithm for Europe. Energy Build. **34**(6), 623–635 (2002)
54. Almeida, R.M.S.F.: Avaliação do Desempenho Higrotérmico do Parque Escolar Reabilitado. PhD Thesis, Faculdade de Engenharia da Universidade do Porto (2009)
55. Curado, A.J.C.: Thermal comfort and energy efficiency in social housing buildings rehabilitated. PhD Thesis in Civil Engineering, Faculty of engineering of the University of Porto, Porto (2014)
56. Gratia, E., Bruyère, I., De Herde, A.: How to use natural ventilation to cool narrow office buildings. Build. Environ. **39**(10), 1157–1170 (2004)
57. Santamouris, M.: Energy in the urban built environment. In: Allard, F., Ghiaus, C. (eds.) The Role of Natural Ventilation, Natural Ventilation in the Urban Environment: Assessment and Design. United Kingdom (2005)
58. Schiffer, S.R., Fleet, A.B.: Manual of Thermal Comfort, 7th edn. Studio Nobel, São Paulo (2003)
59. Mazon, C.O., Silva, R.G., Souza, A.H.: Natural ventilation in warehouses: the use of lanterns in the covers. Mag. School Mines. **59**(2)
60. Hunzinker, D.V.: Study of the phenomena of natural ventilation in buildings. In: 9th Congress of Internal scientific initiation of Unicamp, Campinas-SP (2001)
61. Jerónimo, R.M.S.: Hygrothermal performance assessment and the comfort of Rural Buildings Rehabilitated. PhD thesis in Civil Engineering, Faculty of Engineering of the University of Porto (2012)
62. Gandemer, J., Barnaud, G.: Ventilation naturelle des habitations sous climat tropical humide: aerodynamique aproach. Report for the CSTB, Nantes (1989)
63. Huet, O., Celaire, R.: Bioclimatisme en zone tropicale: construire avec le climate. Groupe de Recherche et d ' Techonologiques Exchanges-GRET. Ministere de la Cooperation. Programme Rexcoop Interministeriel, Paris (1986)
64. Andreasi, W.A.: Natural Ventilation as a Strategy Aiming to Provide Thermal Comfort and Energy Efficiency in the Internal Environment. UFMS (2007)
65. Nicol, F.: Adaptive thermal comfort standards in the hot-humid tropics. Energy Build. **36** (2004)
66. Shaviv, E., Yezioro, A., Capeluto, I.G.: Thermal Mass and Night Ventilation the Passive Cooling Design Strategy. Israel Institute of Technology, Faculty of Architecture and Planning of Haifa, Haifa, Israel (2001)
67. Santamouris, M., Sfakianaki, A., Pavlou, K.: On the efficiency of night ventilation, techniques applied to residential buildings. Environmental Studies Group of buildings, Department of applied physics, University of Athens, Athens, Greece (2010)
68. Almeida, R.M.S.F., de Freitas, V.P., Delgado, J.M.P.Q., Paula, P.: Hygrothermal performance of a naturally ventilated gypsum house—the Brazilian climate influence. In: Proceedings of the 13th International Conference on Building Materials and Components (XIII DBMC), pp. 297–304, São Paulo, Brazil, 2–5 Sept 2014

69. Paula, P., Vázquez da Silva, M., Delgado, J.M.P.Q.: Numerical analysis of hygrothermal building performance of gypsum houses in different Brazilian climates. Diffus. Found. **10**, 132–148 (2017)
70. Pires Sobrinho, C.W.A.: Vertical seals in masonry of blocks of Plaster for Aporticadas reinforced concrete Structures-design, implementation and performance. Technical Document, Recife (2009)
71. Lordsleem, B.C. Jr.: Constructive method of vertical seal with plaster blocks. Research Project, FACEPE, Recife (2009)
72. BEN-2013 National Energy Balance.: Year 2012 Base/Energy Research Company. EPE, Rio de Janeiro (2013)
73. Lyra Sobrinho, A.C.P., Amaral, A.J.R., Dantas, J.O.C.: Gypsum. Mineral contents DNPM, p. 175–178 (2006)
74. SINDUSGESSO.: Cadbury Pole. Available at http://www.sindusgesso.org.br. Access in 02 April 2016
75. Luz, A.B., Lins, F.F.: Industrial minerals & rocks—CETEM-MCT-Mineral Technology Center—Ministry of Science and Technology, Rio de Janeiro (2005)
76. Mineral Contents, DNPM/PE (2006)
77. Mancino, N.: Gypsum in the Middle East. From antiquity to modern day. Gypsum Global Magazine, May 2008
78. Pires, C.W., Balogun, N.M., Costa, T.C., Silva, C.B.: Vertical Fences in masonry of blocks of plaster for Aporticadas reinforced concrete structures-design, implementation and performance. In: 52nd Brazilian Concrete, Fortress (2010)
79. Santos, A.: Hygrothermal behavior of plaster walls-fitness for climatic zones of Brazil, flat. Thesis Submitted for partial satisfaction of the requirements of the Degree of doctor of Civil Engineering, Faculty of Engineering of the University of Porto (2014)
80. Abeysundara Bael, G.Y., Sgheewala, S.A.: Matrix in life cycle perspective is selecting sustainable materials for buildings in Sri Lanka. Build. Environ. **44** (2008)
81. Tavares, S.F.: Methodology for life cycle energy analysis of residential buildings. PhD Thesis in Civil Engineering, UFSC. Florianopolis (2006)
82. Sayil, B., Çolak, A.:Drability of polimer modified and impregnated gypsum. Durability of building materials and components. In: Lacasse, M.A., Vanier, D.J. (eds.) Institute for Research in Construction, Ottawa ON, K1A 0R6, Canada, pp. 485–495 (1999)
83. Okino, E., de Souza, M.R., Santana, M.A.E.: da Alves, M.V., de Sousa, M.E., Teixeira, D. E.: Cement-bonded Wood particleboard with a mixture of eucalypt and rubberwood. Cem. Concr. Comp. **26**(6), 729–734 (2004)
84. Okino, E., de Souza, M.R., Santana, M.A.E.: da Alves, M.V., de Sousa, M.E., Teixeira, D. E.: Physico-mechanical properties and decay resitance of *Cupressus* spp. Cement-bonded particleboards. Cem. Concr. Comp. **27**(3), 333–338 (2005)
85. Frick, E.: The bison system for the production of wood gypsum particleboards. In: Moslemi, A.A, Hamel, M.P. (eds.) International Conference on Fiber and Particleboard Bonded with Inorganic Binder, Idaho, USA, pp. 98–102 (1988)
86. Deng, Y., Xuan, L., Feng, Q.: Effect of a waterproof agent on gypsum particleboard properties. Holzforschung **60**(3), 318–321 (2006)
87. Deng, Y., Furuno, T. (2001), Properties of gypsum particleboard reinforced with polypropylene fiber. J. Wood Sci. **47**(4), 445–450
88. Deng, Y., Furuno, T.: Study on gypsum—bonded particleboard reinforced with jute fibers. Holzforschung **56**(4), 440–445 (2002)
89. Feng-Ging, Z., Hong-jie, Liu, Li-xia, Hao, Qian, Li: Water resistant block from desulfurization gypsum. Constr. Build. Mater. **27**, 531–533 (2012)
90. Bülichen, D., Plank, J.: Water retention capacity and working mechanism of methyl hydroxypropyl cellulose (MHPC) in gypsum plaster which impact has sulfate? Cem. Concr. Res. **46**, 66–72 (2013)
91. Peres, L., Benachour, M., Santos, V.A.: Gypsum Plaster: Production and Use in Construction. Bagaço, Recife (2001)

92. Baltar, C.A.M., Bastos, F.F., Luz, A.B.: Gipsita—LEP. Mineralogical varieties and processes used in the production of different types of gypsum. In.: Encontro Nacional de Tratamento de Minéros Metalurgia Extrativa, Anais. Florianópolis, Brazil. (in Portuguese) (2003)
93. Araújo, S.M.S.: The Araripe gypsum pole: geo-environmental units and mining impacts. PhD thesis, Campinas, Universidade Estadual de Campinas, Brazil (2004)
94. Pinheiro, S.M.: Recycled gypsum plaster: properties evaluation for components use. PhD thesis, Campinas, UNICAMP, Brazil (2011)
95. Nolhier, M.: Construire en platrê. L'harmattan, Paris (1986)
96. Asociación del Yeso, Business Execution and Technique of Covering with Plaster. Available at http://www.atedy.es/buscarPub.asp. Access in 17 Oct 2016
97. Les Industries du Platrê.: Les atouts du plâtre. Available at http://www.lesindustriesduplatre.org/docs/atouts.pdf. Accessed on 10 Apr 2017
98. Brazilian Gypsum.: www.braziliangypsum.com/noticias (2008). Access in 12 July 2017
99. UNE EN 12859, Gypsum blocks-Definitions, requirements and test methods, Brazil, 2008
100. UNE EN 12860.: Gypsum based adhesives for gypsum blocks-Definitions, requirements and test methods, Brazil (2001)
101. UNE EN 15318.: Design and application of gypsum blocks, Brazil (2007)
102. UNE EN 13279-1.: Gypsum binders and gypsum plasters, Brazil (2008)
103. NF P 72-202.: verticaux de plâterie Ouvrages NE nécessitant shovels 1 ' application d'un enduit au plâtre-Exécution des cloisons en carreaux de plâtre, Association Française de Normalisation, Paris (1994)

Influence of Reinforced Mortar Coatings on the Compressive Strength of Masonry Prisms

A. C. Azevedo, João M. P. Q. Delgado and A. S. Guimarães

Abstract This work describes an experimental study is carried out on running bond 195 red clay prisms, of two and three ceramic blocks, with and without cement mortar coating, subjected to axial compression in order to enhance the capacity of masonry. The prisms were subjected to compressive loading and all of them had deformation control on each face with a deflectometer, in order to obtain information about the behaviour of the prisms. The experimental results indicate an increase both in the compressive load capacity of the coated prisms and in those that use coatings based on reinforced mortar, not complying with the specifications of conventional structural mortar. The load ratios of prisms/wallettes and prisms with two blocks/prisms with three blocks were satisfactory.

Keywords Prisms · Masonry · Reinforced mortar · Wallettes

1 Introduction

It is important in masonry design to determine the appropriate ultimate compressive strength of the masonry material. Masonry is a material built from units and mortar that induce an anisotropic behaviour for the composite. The lack of knowledge on the properties of the composite material imposes low assessments of the strength capacity of the masonry wall [1]. Atkinson et al. [2] state that the prediction of compressive strength and deformation of full scale masonry based on compressive tests of stack-bond masonry prism and the interpretation of the results of prism tests

A. C. Azevedo · J. M. P. Q. Delgado (✉) · A. S. Guimarães
CONSTRUCT-LFC, Civil Engineering Department,
Faculty of Engineering, University of Porto, Porto, Portugal
e-mail: jdelgado@fe.up.pt

A. C. Azevedo
e-mail: Antonio.costaazevdo@fe.up.pt

A. S. Guimarães
e-mail: anasofia@fe.up.pt

have a significant influence on the allowable stress and stiffness used in masonry design.

There have been numerous studies done on the behaviour of masonry prisms under axial compression. The effects of variables such as the height-to-thickness ratio of the prism, mortar type and grout strength, unit geometry, and various capping compounds have been the point of focus of many researchers. However, most of the research reports have been presented and published in various conferences around the world; however, some are unpublished [3]. Moreover, some of the data that is the basis of the formula, the graphs, and the design tables presented in various parts of the Masonry Standards Joint Committee specification were the result of research done by the former Brick Institute of America, now the Brick Industry Association.

Structural masonry using hollow clay blocks has begun in Brazil in the mid 80's in residential constructions up to 6 storey high buildings. In the last two decades clay structural masonry system has increased in some states of Brazil, mainly by the availability of blocks of high compressive strength produced in modular sizes. However, tests carried out in the last twenty years in Brazil [4] have shown that the average compressive strength of unreinforced clay walls is only 34% of the average compressive strength of the blocks, and the average compressive strength of two block prisms is 50% of the unit strength [5].

Some structural clay block producers may provide blocks of two or three resistances, with different prices, depending on composition of clay mix, burning temperature, and even different cross sections varying coring patterns.

Despite the great interest, only few studies have been carried out and published in Brazil on the influence of reinforced mortar coating on the compressive strength of clay brick masonry prisms.

2   Materials and Methods

2.1   Geometrical Characteristics

The geometrical characteristics of the clay blocks, i.e., their shape and manufacturing dimensions, must meet the tolerances provided:

- Face measures—Effective dimensions;
- Thickness of the septa and external walls of the blocks;
- Deviation from the square (D);
- Face plane (F);
- Gross area (Ab).

The apparatus required to carry out the measurements consisted of: pachymeter with a minimum sensitivity of 0.05 mm, a metal ruler with a minimum sensitivity of 0.5 mm, a metal bracket of 90 + 0.5° and a balance with a resolution of 10 g, all

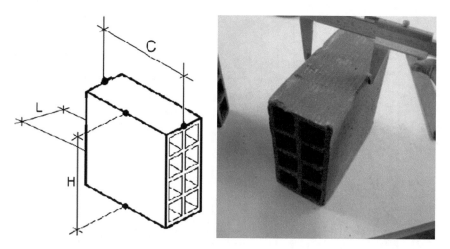

Fig. 1 Septa measurement

of them were properly calibrated. The measurements of the block faces, the values of width (W), height (H) and length (L) were determined as shown in Fig. 1.

The septa measurement had the following procedure and the measurements were made in the central region of the clay blocks, using at least four measurements, searching for the narrowest septa, as we Shown in Fig. 2a. The flatness of the faces was determined by the arrow formed in the diagonal of one of the facing faces of the block, according to Fig. 2b.

Finally, the value of the gross area of each clay block was determined as shown in Table 1, together with all the data of the geometric characterization.

(a)   **(b)**

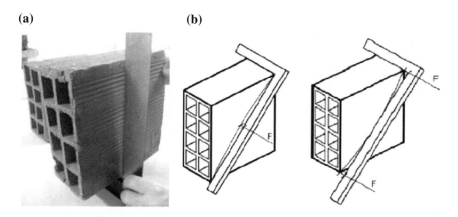

Fig. 2 a Measurement of the deviations and **b** planar measuring. *Source* NBR 15270-3 [6]

Table 1 Geometric characterization of the ceramic blocks tested

Specimens		Face measures—effective dimensions (mm)			Thickness of the septa and external walls (mm)		Square deviation (mm)	Face plane (mm)	Gross area (cm²)
		Width (W)	Height (H)	Length (L)	Septa	External walls	D	F	Ab
1		93	193	190	7	8	1	1	177
2		91	191	191	7	8	1	1	173
3		93	192	190	8	8	1	1	176
4		90	192	190	8	8	1	1	171
5		91	190	188	8	8	1	1	172
6		91	190	191	7	8	1	1	175
7		91	191	189	8	9	1	1	173
8		90	190	191	8	8	1	1	172
9		92	188	190	7	8	1	1	174
10		92	192	191	8	9	1	1	176
11		92	191	192	8	8	1	1	175
12		89	190	188	7	8	1	1	167
13		91	189	193	7	8	1	1	175
Average		91	191	190	8	8	1	1	174
Tolerance NBR 15270-1 [19] (mm)	Individual ±	5	5	5	Minimum 6	Minimum 7	Maximum 3	Maximum 3	x
	Average ±	3	3	3					
Non-conforming units		0	0	0	0	0	0	0	x

Reference dimensions of the clay blocks: 90 × 190 × 190 mm

2.2 *Physical Characteristics*

The physical characteristics of the ceramic blocks analysed were:

- Dry mass (m_{dry});
- Water absorption index (AA);
- Initial absorption index (AAI).

The equipment used for such determinations were: balance with a resolution of up to 5 g, oven with adjustable temperature and a tank for submersion of the samples. The number of samples established by the standard to determine these characteristics is equal to six specimens.

To determine the dry mass (m_{dry}) the samples of the ceramic blocks were first cleaned for dust removal and other loose particles adhered to the block and identified, after which they were submitted to a temperature range of $(105 + 5)$ °C up to the stabilization of the individual mass, when after two consecutive weighing, with intervals of 1 h, did not differ by more than 0.25% of the weight, as shown in Fig. 3a.

After the determination of the dry mass, the blocks were completely submerged in a tank with water at room temperature for a period of 24 h and then removed and placed in a balance to determine the wet mass (m_{wet}) after removal of excess water with a cloth (see Fig. 3b).

The water absorption index (AA) of each specimen was calculated using the following expression:

$$AA(\%) = \left[\left(m_{wet} - m_{dry} \right) + m_{dry} \right] \times 100 \tag{1}$$

The determination of the initial absorption index (IRA) required, in addition to the equipment described above, a chronometer with a sensitivity of 1 s, a bubble level ruler, a water reservoir that allows the maintenance of a $(3 + 0.2)$ mm as shown schematically in Fig. 4.

The samples were first subjected to heating in an oven for 24 h and after their removal 2 h were allowed for cooling in the open air until the ambient temperature. The geometric characteristics of the blocks were determined to obtain the contact area with the water slide. The blocks were then positioned on the supports so that the contact face of the block remained in contact with the water slide at a height of 3 mm for a time of 1 min, then the block was removed and removed the excess water with the aid of a damp cloth to proceed with the weighing of the block.

The IRA is the water absorption index (suction) of the tested face of the blocks, expressed in $(g/193.55 \ cm^2)/min$ and is calculated according to the expression:

$$IRA = 19.55 \times (\Delta P + area) \tag{2}$$

(a)   **(b)**

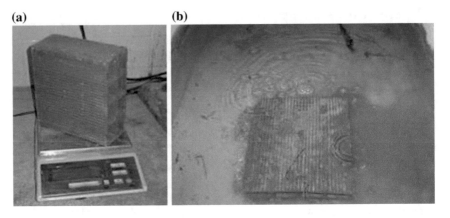

Fig. 3 a Wet weight and **b** submerged specimen

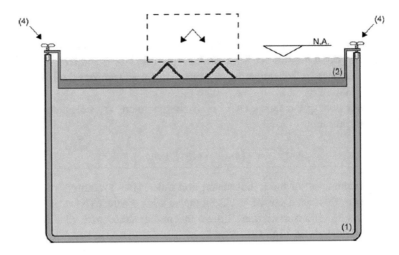

Fig. 4 Sketch of IRA determination. *Source* NBR 15270-3 [6]

where ΔP is the wet mass change after 1 min (dried at room temperature). The properties of the clay bricks used in prism construction are presented in Table 2. All the samples used in testing had a net area that exceeded 75% of their gross area.

2.3   Mechanical Characteristics

The mechanical characteristics of the ceramic clay brick blocks were evaluated by individual compressive strength. For the accomplishment of this test a total of 13 specimens were properly prepared. The regularization of the two faces destined to

Table 2 Properties of the clay brick blocks used

Samples	IRA[a]	Dry mass (g)	Compressive strength (MPa)[b]
1	6.9	2406	2.17
2	12.6	2320	2.10
3	9.2	2289	2.16
4	8.6	2338	2.12
5	17.4	2307	2.02
6	3.9	2403	2.04
Average	9.77	2331.4	2.05
Standard deviation	±4.70	±43.86	±0.40
COV[c]			19%

[a] IRA: Initial Rate of Absorption, expressed in grams per minute per 30 in^2 (193.55 cm^2)
[b] Compressive Strength: Calculated using the gross area of the unit, MPa
[c] COV: Coefficient of Variation in %

the settlement perpendicular to the block length, was done with cement and a maximum thickness of 3 mm in order to uniformity the block surfaces.

After hardening the capping layers, the specimens were completely submerged in a tank with water for a period not lesser than 6 h, as established in the Brazilian standard NBR 15270-3 [6]. The machine used to carry out the tests was the Universal Testing Machine a facility of the Laboratory of Construction Materials of Catholic University of Pernambuco, Brazil. The specimens were tested in saturated conditions and placed in the press so that their center of gravity coincides with the load axis of the press plates, as illustrates in Fig. 5. The results obtained are presented in Table 1.

(a) **(b)**

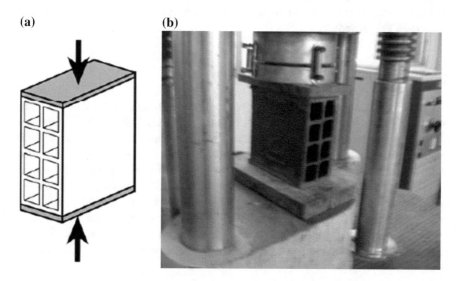

Fig. 5 **a** Load application axis and **b** specimens before the experimental tests

2.4 Fine Aggregate Analyse

The characterization of the fine aggregates (sand), used in the preparation of mortars for coating the prisms, took into account the following tests and their respective Brazilian standards:

- Granulometry of the finr aggregate—NBR NM 248 [7]
- Specific Gravity Flask of Chapman—NBR 9776 [8]
- Clay content in clods—NBR 7218 [9]
- Determination of fine materials—NBR NM 46 [10]
- Determination of the unit mass—NBR NM 45 [11]
- Swelling of the fine aggregate—NBR 6467 [12].

The sand used was acquired in the Metropolitan Region of Recife and all the tests were carried out in the Laboratory of Construction Materials of Catholic University of Pernambuco, Brazil.

2.4.1 Determination of the Granulometric Composition

The determination of the granulometric composition aims to classify the aggregate as a function of the size of the grains. The test method is described in NBR NM 248 [7] and to determine if it is necessary to collect two samples of the aggregate to be analysed which must then be washed and preheated to a temperature of $(105 + 5)$ °C. Its classification occurs through a set of sieves with openings standardized by ABNT (see Fig. 6). The assembly is organized in a decreasing manner so that the sieves of larger apertures overlap the sieves of smaller apertures; the material is then sieved so that each fraction is retained in the sieves, and then separated and weighed. Figure 6 illustrates the procedure.

2.4.2 Determination of the Specific Mass

In order to determine the specific mass of the fine aggregate, the procedure described in the Brazilian standard NBR NM 52 [13], using the Chapman vial (see Fig. 7), a standardized vial that indicate the displacement of the water column volume after the materials insertion, was used. The material specific mass is the ratio of its mass and the volume occupied by it, excluding voids between the grains.

The sand was oven dried at a temperature of about 105 °C and a sample of 500 g was withdrawn. This sample was then inserted into the Chapman bottle, which was already filled with water until a volume of 200 cm^3; as the material was being placed, the water column moved so that at the end, the variation of this displacement represented the volume of sand inserted. The equation used is shown below:

Fig. 6 Sieves used

$$\gamma = m \div (L - 200) \tag{3}$$

where m is the dry mass of the material and L is the final volume of the water column, the result is expressed in g/cm^3.

2.4.3 Determination of the Fine Materials Content

Fine materials present in the aggregate, i.e., those passing through the 75 μm aperture sieve, also called powder material, should be analysed by the method described in NBR NM 46 [10]. Its quantity, when higher than the one foreseen in norm NBR 7211 [14] that is of 5%, can harm the mixture, either of concrete or mortar, because the very fine grains make difficult the adhesion of the paste of cement to the aggregate.

The sample was oven-dried at 110 °C until mass constancy and then a 100 g sample was withdrawn. This quantity was placed on the sieves with opening 1.2 mm and 75 μm and subjected to the washing successive times so that the fine material adhered to the aggregate was eliminated with the water. The process was completed when the water passed through was completely clean. The material was again placed in an oven to evaporate the water and obtain the final mass. The result was calculated according to the following expression:

Fig. 7 Chapman bottle

$$\text{Powder material content} = \left[\left(M_i - M_f\right) \div M_i\right] \times 100 \qquad (4)$$

where M_i is the initial mass and M_f is the final mass.

2.4.4 Clay Content in Clods

Clad clays and friable materials are materials that are susceptible to wear when subjected to minor stresses that may alter the quality of an aggregate for contamination with poorly resistant grains and which impair both the strength and the appearance of concrete and mortars. Its content is calculated according to the recommendations of norm NBR 7218 [9], separating a portion of the kid's aggregate that passes in the sieve with opening 4.8 mm and is retained in the sieve with opening 1.2 mm, identifying the clods and friable grains and proceeding with discharging of these grains and subjecting them to a new sifting process. The calculation is based on the following equation:

$$\text{depleted material} = (m_i - m_f)/m_i \qquad (5)$$

where m_i is the initial mass and m_f is the final mass after sieving.

2.4.5 Unit Mass Test

The unit mass in the loose state of the fine aggregate is determined by NBR NM 45 [11]. In this method a cylindrical vessel with known volume of 20 l, 16 mm diameter metal rod, a concrete shovel, metal ruler, a balance with resolution of 50 g and an amount of dry sand sufficient to occupy the volume of the vessel.

In this test, method C of the standard was used. The dry sand is added without compaction until the entire volume of the vessel is occupied, the metal ruler is used to remove the excess sand on the vessel and the sand set plus vessel is weighed, the result is calculated according to the expression:

$$\rho_{ap} = (m_{ar} - m_r) \div V \qquad (6)$$

where ρ_{ap} is the unit mass of the aggregate (kg/m^3), m_{ar} is the mass of sample plus container (kg), m_r is the empty container mass (kg) and V is the container volume (m^3).

2.4.6 Swelling of the Fine Aggregate

The swelling is a phenomenon that concerns the small aggregate and can be described as the variation of the apparent volume that affects the unit mass of the material when it is submitted to the variation of the moisture content. In other words, the same amount of sand may occupy larger or smaller volumes without compaction when its moisture content varies from dry to wet sand. When performing a volume trace, it is very important to correct the volume of the sand in the padola as well as its moisture content, due to the great capacity of water retention by the small aggregates.

The test, according to the standard NBR 6467 [12], is to perform several measurements of the unit mass in various sand moisture conditions, namely: 0, 0.5, 1, 2, 3, 4, 5, 7, 9 and 12% and the coefficient of swelling is then calculated by the expression:

$$CI = (V_h \div V_0) = [(\gamma_s \div \gamma_h) \times ((100 + h) \div 100)] \qquad (7)$$

where V_h is the volume of the aggregate with h% humidity (dm^3), V_0 is the volume of dry aggregate in greenhouse (dm^3), V_h/V_0 is the swelling coefficient, γ_s is the unit mass of the dry aggregate (kg/dm^3), γ_h is the aggregate mass with h% of moisture content (kg/dm^3) and h is the aggregate moisture content (%).

Table 3 Granulometric results of the fine aggregates [7]

Opening of the sieves (mm)	Maas retained (%)		Variation of the % retained ≤ 4 (%)	Average of retained mass (%)	Cumulated of retained mass (%)	Ratios over % retained			
						Lower limits		Higher limits	
	Test no 1 (%)	Test no 2 (%)				Used zone	Optimal zone	Used zone	Optimal zone
9.5	0.0	0.0	0.0	0.0	0.0	0	0	0	0
6.3	0.0	0.0	0.0	0.0	0.0	0	0	0	7
4.75	0.6	0.3	0.2	0.5	0.5	0	0	5	10
2.36	4.1	3.3	0.8	3.7	4.1	0	10	20	25
1.18	10.1	8.9	1.2	9.5	13.6	5	20	30	50
0.6	21.4	21.4	0.0	21.4	35.1	15	35	55	70
0.3	33.4	32.5	0.9	33.0	68.0	50	65	85	95
0.15	21.2	23.1	1.9	22.1	90.2	85	90	95	95
Bottom	9.2	10.5	1.2	Fine module = 2.11		Maximum characteristic size (mm) = 2.4			

Table 4 Summary of the characterization results obtained with the fine aggregates

Specific mass (g/cm^3)	Clay clods (%)	Fine material (%)	Unitary dry mass (kg/m^3)	Swelling	
				Swelling (%)	Critical humidity (%)
2.62	0.0	4.4	1450	1.23	3.20
NBR 9776	NBR 7218	NBR NM 46	NBRNM45	NBR 6467	

Finally, Table 3 shows the granulometric results according to the Brazilian standard NBR NM 248 [7] and Table 4 presents a presents a summary of the characterization results obtained with the fine aggregates used in the research.

2.5　Weak, Medium and Strong Mortar

Experimental tests with red clay prisms were performed in the Laboratory of Construction Materials of Catholic University of Pernambuco, Brazil. The red clay brick blocks used had dimensions of 91 × 191 × 190 mm^3 (with a tolerance of ± 3 mm) and an average density of 2620 kg/m^3. Overall, a total of 195 prisms were built and tested (see Table 5). All the applicable ASTM standards or Brazilian standards (NBR) were followed in building, curing, capping, and testing of the prisms and the components.

3　Results and Discussion

3.1　Mortar

The mortars used in this work were subject to a detailed analysis process in which their properties were investigated in the fresh state as in the hardened state. Table 6 shows the grout mixtures (mix ratio) studied as well as their applications.

The fresh mortars were characterized by consistency and density test. The mortar consistency was evaluated through the procedures described in the Brazilian standard NBR 7215 [15], allowing to verify the plasticity degree with a Flow Table. The samples were subjected to successive fall from a pre-established height, so that the more plastic the mortar, the greater its final diameter.

A mortar may be considered dry when the consistency index (flow table) is less than 250 mm. Mortars with a consistency index between 260 and 300 mm (ex. plaster mortar) are considered plastic. Finally, mortars with a consistency index of more than 360 mm are considered to be fluid. The results are presented in Table 3.

The mortars in the hardened state were characterized by the following tests:

Table 5 Description of the tested samples

Ref.	Samples tested	Dimension (cm × cm)	Area (cm^2)
Prims with 2 clay blocks			
2P-1	Uncoated prims	9 × 19	171
2P-3	Prisms with a coating of 3 cm of mix ratio 1:1:6 (cement: lime:sand)	15 × 19	285
2P-5	Prisms with a coating of 3 cm and mix ratio 1:2:9 (cement:lime:sand)	15 × 19	285
2P-7	Prisms with a coating of 3 cm and mix ratio 1:1:6 (cement:lime:sand) reinforced with a POP mesh of 10 × 10 cm and 4.2 mm of diameter	15 × 19	285
Prims with 3 clay blocks			
3P-1	Uncoated prisms	9 × 19	171
3P-2	Prisms with a mix ratio of 1:3 (cement:sand)	10 × 19	190
3P-3	Prisms with a coating of 3 cm of mix ratio 1:1:6 (cement: lime:sand)	15 × 19	285
3P-4	Prisms with a coating of 1.5 cm and mix ratio 1:2:9 (cement:lime:sand)	12 × 19	228
3P-5	Prisms with a coating of 3 cm and mix ratio 1:2:9 (cement:lime:sand)	15 × 19	285
3P-6	Prisms with a coating of 3 cm of mix ratio 1:0.5:4.5 (cement:lime:sand)	15 × 19	285
3P-7	Prisms with a coating of 3 cm and mix ratio 1:1:6 (cement:lime:sand) reinforced with a POP mesh of 10 × 10 cm and 4.2 mm of diameter	21 × 19	399
3P-8	Prisms with a coating of 1.5 cm and mix ratio 1:2:9 (cement:lime:sand) reinforced with a POP mesh of 10 × 10 cm and 4.2 mm of diameter	18 × 19	342
3P-9	Prisms with a coating of 3 cm and mix ratio 1:2:9 (cement:lime:sand) reinforced with a POP mesh of 10 × 10 cm and 4.2 mm of diameter	21 × 19	399

- Axial compressive strength (rupture test), NBR 5738 [16];
- Diametrical compressive tensile strength, NBR 7222 [17].

In order to measure the axial compressive strength, six cylindrical specimens were prepared for each mortar type, with diameter of 5 cm and a height of 10 cm. The specimens were cured for a period of 28 days, before carrying out the resistance tests, as showed in Fig. 8 and described in Table 6.

In order to perform diametrical compressive tensile strength tests, the procedures described in the Brazilian standard NBR 7222 [17] were used, which establishes the preparation of six specimens, in the same way in which they were prepared for the

Table 6 Consistency index, axial compression strength and tensile strength by diametrical compression of the mortars tested

Mortar application	Grout mixtures (mix ratio)	Consistency index (mm)	Axial compressive strength (kN)	Tensile strength (MPa)
Settlement	1:1:6 (cement: lime:sand)	266	6.5 ± 1.1	0.9 ± 0.2
Roughcast	1:3 (cement:sand)	305	30.4 ± 0.6	2.8 ± 0.9
Coating	1:2:9 (cement: lime:sand)	302	2.8 ± 0.4	0.8 ± 0.3
Coating	1:1:6 (cement: lime:sand)	306	6.5 ± 1.1	0.9 ± 0.2
Coating	1:0.5:4.5 (cement: lime:sand)	296	5.4 ± 0.5	0.9 ± 0.8

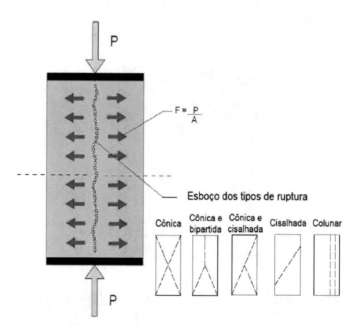

Fig. 8 Sketch of the rupture test

axial compression resistance tests. The specimens were positioned in order that the contact between the plates of the test machine and the specimens gives only two generators diametrically opposite to the specimen. The results obtained are presented in Table 6.

3.2 Prisms and Wallettes

In order to perform the mechanical deflectometer readings, 400 L-shaped metallic plates, with 6 cm high and 2 cm wide, were fabricated (see Fig. 9), which were used as support base for the deflectometers. The metallic plates were fixed to the prisms, previously, in the middle third of their length, through bonding, at a distance sufficient to allow the free flow of the deflectometers, as seen in Fig. 9a.

The hydraulic jacks had 200 mm piston stroke and 50 ton load capacity. This allowed only one jack to be sufficient to apply the load required for rupture of the prisms. The load drive machine, controlled by software which allows a perfect control of both displacement increment and load increase, has a servo-hydraulic working system and it is connected to the linear displacement sensors (LVDT). The displacement control of the hydraulic jacks makes it possible not only to follow the post-cracking and post-rupture, but also the shape of the rehab curve of the samples in front of the maintenance or increase of displacement.

(a)

(b) **(c)**

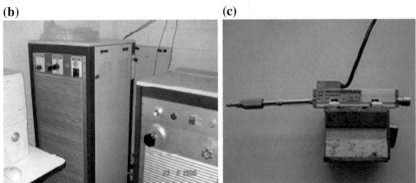

Fig. 9 a Deflectometer; **b** monitors drive machine and **c** LVDT (linear variable differential transformer)

In this research we promote two distinct, but simultaneous, ways of measuring the displacements. The first one used two displacement sensors, or LVDT's (Linear Variable Differential Transformer), a particular type of variable reluctance sensor (see Fig. 9c). The second way of displacement measurement, came from the need to compare the measurements obtained with LVDT's, which take into account the entire length of the prism along with the wood deformation, with measurements of section parts of the prism length. Several procedures were tested and it was decided to measure the middle third. In this way, two pre-bonded metal plates were placed on each side of the prism to serve as support for the mechanical extensometers.

Table 7 shows that the average compressive resistance of the uncoated prisms with two (2P-1) or three blocks (3P-1) was not very different. In fact, for the two-block prism, the mean load was 9.72 kN whereas for the three-block prisms this load was 9.49 kN, representing a difference of 2.4%. However, it is important to note that the coefficients of variation observed were 35 and 22%, for uncoated prisms with two and three blocks respectively, indicate that the observed values should be examined with caution. In addition, it should be noted that these prisms were tested on a hydraulic press machine with a load capacity of 500 kN and the average compressive resistance values obtained were below the accuracy of the press machine, which is on the order of 10% of its capacity. In spite of these aspects it is possible to conclude that no significant difference was observed between prisms of two and three blocks.

Coated prisms made with weak and medium mortar, with 2 and 3 blocks, according to Table 7, had a significant increase of load capacity, reaching up to ∼420% for 2 block prisms and ∼450% for 3 block prisms.

Table 7 Prisms and wallettes test results

Sample	Prisms compressive resistance (kN)			Wallettes compressive resistance (kN)		
	Average	St. Dev.	COV (%)	Average	St. Dev.	COV (%)
2P-1	9.72	3.38	35	56.3	8.7	15.4
2P-3	45.47	12.21	27	168.3	33.3	19.8
2P-5	50.43	13.61	27	156.5	16.1	10.3
2P-7	120.73	16.89	14	417.1	63.0	15.1
3P-1	9.49	2.08	22	56.3	8.7	15.4
3P-2	18.05	5.92	33	84.9	16.3	19.2
3P-3	52.71	9.07	17	168.3	33.3	19.8
3P-4	39.58	9.22	24	130.4	24.4	18.7
3P-5	45.03	10.38	23	156.5	16.1	10.3
3P-6	59.02	8.81	15	262.2	42.7	16.3
3P-7	109.17	11.23	10	417.1	63.0	15.1
3P-8	94.47	12.17	13	321.0	47.7	14.9
3P-9	100.25	10.54	11	367.0	49.3	13.4

For the 2 block prisms with coating, it was observed that the prisms with medium mortar presented lower compressive load capacity than the prisms made with weak mortar, although the difference did not exceed 10%. The coefficient of variation observed for these two situations was of the same order of magnitude, approximately 27%, being this value considered statistically high, possibly explaining the unexpected result.

The observed inconsistency in the mean burst load of the 2 block prisms was not observed in the 3 block prisms. This can be justified by the fact that the coefficient of variation for the 3 block prisms was lower than for 2 block prisms.

The grout mixtures and the thickness of the mortar in the 3 block prisms had a significant influence on the compressive resistance of these elements. It is possible to observe in prisms with a single coating mortar (1:1:6), that increasing the thickness from 1.5 to 3 cm the load capacity increased approximately 14%. While for prisms with the same thickness of 3 cm and different mix ratio it was observed and increase of 31% between mix ratios of 1:1:6 and 1:0.5:4.5.

Reinforced prisms with 3 cm of thickness reinforced mortar and connectors showed a significant increase in the load capacity when compared to the prisms without framework and connectors. For the prisms with 2 blocks and a medium mortar with 3 cm of thickness the observed increase was 165%. For the 3 block

Table 8 Prism load ratios

Typology	3-blocks prisms/ wallettes	Relation of areas	2-blocks prisms/ wallettes	Relation of areas	3-blocks/ 2-blocks prisms
Uncoated prisms	0.17	0.53	0.17	0.53	0.98
Prisms with mix ratio 1:3	0.21	0.65	X	X	X
Prisms with a coating of 1.5 cm and mix ratio 1:2:9	0.30	0.93	X	X	X
Prisms with a coating of 3 cm and mix ratio 1:2:9	0.29	0.90	0.32	0.99	0.89
Prisms with a coating of 3 cm of mix ratio 1:1:6	0.31	0.96	0.27	0.84	1.16
Prisms with a coating of 3 cm of mix ratio 1:0.5:4.5	0.23	0.71	X	X	X
Prisms with a coating of 1.5 cm and mix ratio 1:2:9 reinforced.	0.29	0.90	X	X	X
Prisms with a coating of 3 cm and mix ratio 1:2:9 reinforced	0.27	0.84	X	X	X
Prisms with a coating of 3 cm and mix ratio 1:1:6 reinforced	0.26	0.81	0.29	0.90	0.90
Average	0.26	0.81	0.26	0.81	0.98
Standard deviation	0.05	0.16	0.06	0.19	0.12
COV (%)	19	19	24	24	13

(a)

(b)

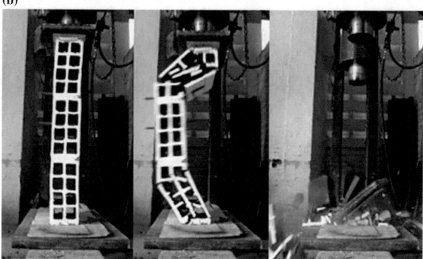

Fig. 10 Rupture of the **a** 2 block prisms and **b** 3 block prisms without coating

prisms with weak mortar, 1.5 or 3 cm thickness, this increase was 139 and 123%, respectively. For the 3 block prisms with medium mortar, 3 cm thickness, the increase was 107%.

Tables 7 and 8 also present a comparison between the average compressive resistance of prisms with 2 and 3 blocks and ceramic wallettes. It is worth noting that the load application area of the prism is 19×9 cm^2 and for the wallettes is 59×9 cm^2, which implies a wall/prism area ratio of 3.105. All the elements tested were made under the same conditions and using the same types of materials and labour.

The coefficients of variation of the rupture load of the reinforced prisms and reinforced wallettes were very similar and relatively low, showing a greater uniformity of the final load. This fact can be explained by the presence of the steel meshes interlocked by connectors inside the mortars.

According to the Brazilian standard NBR 10837 [18], the calculation for the admissible load of a masonry wall, P_{Wall}, is given by:

$$P_{Wall} = 0.20P_{Prism}\left[1 - \left(\frac{h}{40.t}\right)^3\right] \tag{8}$$

where P_{Prism} is the prism rupture load, h is the prism height and t is the prism thickness.

(a)

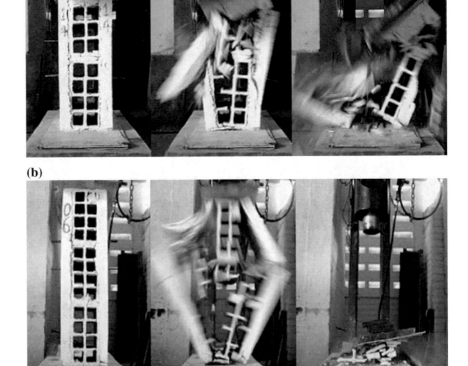

(b)

Fig. 11 Rupture of the **a** 2 block prisms and **b** 3 block prisms with a coating of 3.0 cm and a mix ratio of 1:2:9

In order to obtain a relation between the P_{Wall} and P_{Prism}, not taking into account the safety coefficient and considering the length relation of the wallettes/prism equal to $59/19 = 3.105$, the following relation was obtained: $P_{Wall}/P_{Prism} = 0.344$.

On the other hand, a mean load ratio of the two and three block prisms was obtained experimentally when compared to the results of the wallettes being of the same order of magnitude: 0.26. This behaviour shows that there was no significant difference in the relation between the prisms with 2 or 3 blocks and the wallettes.

(a)

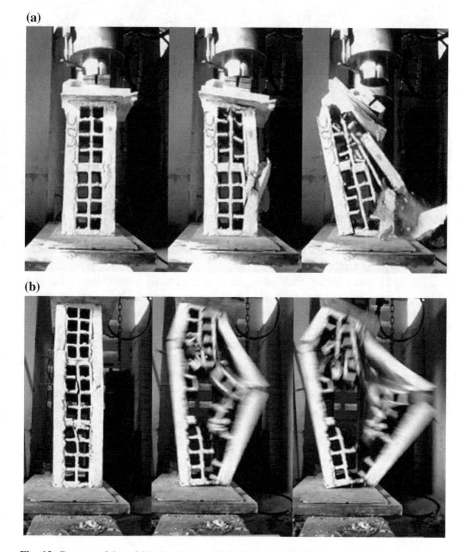

(b)

Fig. 12 Rupture of the **a** 2 block prisms and **b** 3 block prisms with a coating of 3.0 cm and a mix ratio of 1:1:6

Fig. 13 Rupture of the 3 block prisms with a mix ratio of 1:3

The coefficients of variation associate to the rupture load of the reinforced prisms and reinforced wallettes were very similar and relatively low, showing a greater uniformity of the final load. This fact can be explained by the presence of the steel meshes interlocked by connectors inside the mortars.

Figure 10 show the usual type of rupture observed with uncoated prisms, which was abrupt. It is also possible to observe that the prims with 2 or 3 blocks presents the same type of rupture.

Figures 11 and 12 show that the ruptures of the coated prisms start at the septa and it was transferred to the coating layer. This type of rupture it was not abrupt but by shear. The coated prisms present the same type of rupture for 2 or 3 blocks (Fig. 13).

Figure 14 shows the rupture type observed in 2 block prisms and 3 block prisms with a coating of 3.0 cm and a mix ratio of 1: 1: 6. It was possible to observe that the rupture in the reinforced prisms was transferred from the septa to the first layer and then to the second layer. This rupture type was less explosive, continuing to be abrupt (Figs. 15, 16 and 17).

Finally, Figs. 18 and 19 show the mean displacements observed during the application of load in the prisms of two and three blocks, respectively. Table 5 presents the individual results of the displacements of each type of prism tested in this research.

From the analyse of these figures it is possible to observe that for the prisms of 2 and 3 blocks, without coating, the hardness was lesser than in the other prisms

(a)

(b)

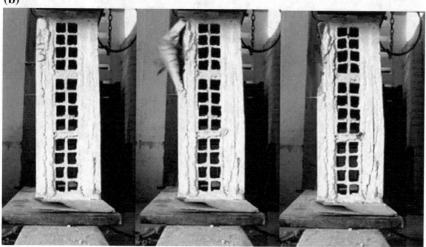

Fig. 14 Rupture of the **a** 2 block prisms and **b** 3 block prisms with a coating of 3.0 cm and a mix ratio of 1:1:6 reinforced with a POP mesh of 10×10 cm and 4.2 mm of diameter

studied. In general, it was not observed significant difference behaviour in the prisms made with strong and weak mortar, both for prisms with 2 and 3 blocks. However, the reinforced prisms present higher performance in comparison with the other prisms studied; aspects that suggest the importance of the reinforcement performed.

Fig. 15 Rupture of the 3 block prisms with a coating of 3.0 cm and a mix ratio of 1:0.5:4.5

Fig. 16 Rupture of the 3 block prisms with a coating of 1.5 cm and a mix ratio of 1:2:9 reinforced with a POP mesh of 10 × 10 cm and 4.2 mm of diameter

Fig. 17 Rupture of the 3 block prisms with a coating of 3 cm and a mix ratio of 1:1:6 reinforced with a POP mesh of 10 × 10 cm and 4.2 mm of diameter

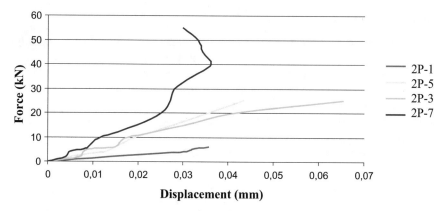

Fig. 18 Load versus displacement for prisms of 2 blocks

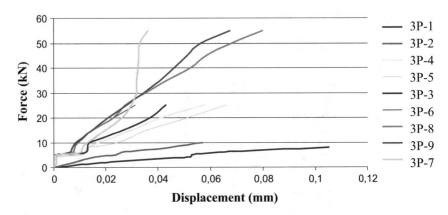

Fig. 19 Load versus displacement for prisms of 3 blocks

4 Conclusions

The main conclusions of this study are as follows:

For prisms with two blocks

- The experimental results showed that the coatings contributed to increase the vertical compressive resistance of the masonry elements studied;
- Several types of rupture were observed in the prisms, and it is not possible to define a typical rupture form. On the other hand, the lateral detachment ruptures of the coating layers were frequent;
- The increase observed in the load related to the reinforcement of the coated prisms was approximately 165%;
- The relation between the maximum loads of rupture of the wallettes and prisms was 0.81.

For prism with 3 blocks

- The coating layer applied on the wallettes of the prisms contributed to increase the vertical load capacity of the studied resistant masonry elements;
- The prisms have an increase of rupture load of more than 210% in relation to the application of a double layer of reinforced mortar with steel mesh;
- The average ratio of the maximum burst loads of the wallettes and prisms was 0.81;
- The ratio of the maximum average loads of 2 block prisms and 3 block prisms was near 1, shown to be equivalent.

References

1. Mohamad, G., Lourenço, P.B., Roman, H.R., Barbosa, C.S., Rizzatti, E.: Stress-strain behaviour of concrete block masonry prisms under compression. In: 15th international brick and block masonry conference, Florianópolis, Brazil (2012)
2. Atkinson, R.H., Noland, J.L., Abrams, D.P., McNary, S.: A deformation failure theory for stack bond, brick masonry prisms in compression. In: Proceedings 3rd NAMC, Arlington, pp. 18-1 to 18-18 (1985)
3. Aryana, S.A.: Statistical analysis of compressive strength of clay brick masonry prisms, M.Sc. Thesis, The University of Texas at Arlington(2006)
4. Cavalheiro, O.P., Gomes, N.S.: Structural masonry of hollow blocks: Test results of elements and compressive strength reducers. In: 7th international seminar on structural masonry for developing countries, proceedings, Belo Horizonte, Brazil, pp. 411–419 (2002)
5. Cavalheiro, O.P., Arantes, C.A.: Influence of grout on hollow clay masonry compressive strength. In: 13th international brick and block masonry conference, Amsterdam, 4–7 July 2004
6. NBR 15270-3: Structural and non-structural ceramic blocks—test methods, Rio de Janeiro, Brazil (2005)
7. NBR NM 248: Aggregates—Sieve analysis of fine and coarse aggregates, Rio de Janeiro, Brazil (2003)
8. NBR 9776: Aggregate—determination of fine aggregate specific gravity by Chapman vessel—method of test, Rio de Janeiro, Brazil (1987)
9. NBR 7218: Aggregates—determination of clay lumps and fiable materials content—method of test, Rio de Janeiro, Brazil (1987)
10. NBR NM 46: Aggregates—determination of material finer than 75 μm sieve by washing, Rio de Janeiro, Brazil (2003)
11. NBR NM 45: Aggregates—determination of the unit weight and air-void contents, Rio de Janeiro, Brazil (2006)
12. NBR 6467: Aggregates—determination of swelling in fine aggregates—method of test, Rio de Janeiro, Brazil (2008)
13. NBR NM 52: Fine aggregate—determination of the bulk specific gravity and apparent specific gravity, Rio de Janeiro, Brazil (2009)
14. NBR 7211: Aggregate for concrete—specification, Rio de Janeiro, Brazil (2009)
15. NBR 7215: Portland cement—determination of compressive strength, Rio de Janeiro, Brazil (1996)
16. NBR 5738: Concrete—procedure for moulding and curing concrete test specimens, Rio de Janeiro, Brazil (2004)
17. NBR 7222: Mortar and concrete—determination of the tension strength by diametrical compression of cylindrical test specimens, Rio de Janeiro, Brazil (1994)
18. NBR 10837: Hollow concrete blocks—bases for design of structural masonry procedure, Rio de Janeiro, Brazil (2000)
19. NBR 15270-1: Ceramic components part 1: hollow ceramic blocks for non-load bearing masonry—terminology and requirements, Rio de Janeiro, Brazil (2005)

Experimental Analyse of the Influence of Different Mortar Rendering Layers in Masonry Buildings

R. A. Oliveira, F. A. Nogueira Silva, C. W. A. Pires Sobrinho,
A. C. Azevedo, João M. P. Q. Delgado and A. S. Q. Guimarães

Abstract This work discusses masonry buildings constructed in the state of Pernambuco, Brazil. Topics such as the main features of this construction technique and the peculiarities that affect its structural behaviour are discussed. Technical information about accidents occurred in recent years are also discussed, along with the historical records of the events, followed by indications of the causes for the collapse. Additionally, this work presents an extensive characterisation of materials and components used in non-structural masonry constructions in the region, making it one of the most comprehensive research studies on this topic in Brazil. This study conducts an in-depth, numerical and experimental analysis of the behaviour of the compressive strength of blocks, prisms and mini-walls that are part of a non-load bearing system, which is often used in the region to carry loads above its own weight. The results obtained allowed to identify the contribution of several mortar rendering layers to the load capacity of the tested specimens. The factors that influenced the load capacity of the tested specimens are also discussed. Finally, a summary of the main results of all the tests performed is presented in order to

R. A. Oliveira · F. A. Nogueira Silva · A. C. Azevedo · J. M. P. Q. Delgado (✉)
A. S. Q. Guimarães
Departamento de Engenharia Civil, Universidade Católica de Pernambuco,
Recife, Brazil
e-mail: jdelgado@fe.up.pt

R. A. Oliveira
e-mail: romildealmeida@gmail.com

F. A. Nogueira Silva
e-mail: farturnog@gmail.com

A. C. Azevedo
e-mail: anasofia@fe.up.pt

A. S. Q. Guimarães
e-mail: Antonio.costaazevedo@fe.up.pt

C. W. A. Pires Sobrinho
CONSTRUCT-LFC, Civil Engineering Department, Faculty of Engineering,
University of Porto, Porto, Portugal
e-mail: carlositep@gmail.com

© Springer International Publishing AG 2018
J. M. P. Q. Delgado and A. G. Barbosa de Lima (eds.), *Transport Phenomena in Multiphase Systems*, Advanced Structured Materials 93,
https://doi.org/10.1007/978-3-319-91062-8_3

provide a detailed explanation for the failures modes observed, which were always sudden and brittle, as was the case in most accidents that occurred with this type of building in the region.

Keywords Masonry buildings · Accidents with masonry buildings
Experimental tests

1 Introduction

There have been numerous studies performed on the behaviour of masonry prisms under axial compression [5, 8, 9]. The effects of variables such as the height-to-thickness ratio of the prism, mortar type and grout strength, unit geometry, and various capping compounds have been the point of focus of many researchers [6, 25].

The occurrence of several accidents in the Metropolitan Region of Recife—MRR—with masonry buildings constructed with non-structural blocks to carry loading beyond its own weight has drawn the attention of the regional and national technical community for the need to establish criteria of research, study and rehabilitation, within acceptable levels of reliability. Masonry buildings constructed with such technique is often referred as resistant masonry buildings (see Fig. 1).

The resistant masonry is one constructive technique characterizes for the use of sealing units (ceramic or concrete) with structural purpose, supporting loads beyond its proper weight. The foundations generally are constructed in masonries with 9 or 19 cm of thickness, in continuity to the walls of the construction, usually seated on low shoes races of armed concrete with transversal section in form of inverted T or on daily pay-moulded components of foundation, seated on layer of concrete regularization.

Diverse pathological manifestations have been observed, already having occurred, in some cases, collapses with fatal victims. It is important to register that the problem in quarrel does not consist in a local exclusiveness, and is of the knowledge of the authors the existence of accident with similar characteristics in Maceió and building pathology manifestations of the same nature in situated building in Belo Horizonte (see Fig. 2).

Approximate numbers indicate that there are around 6000 residential buildings in the region made with this type of masonry buildings where close to 250,000 people live. Several pathological manifestations and collapses with human deaths have also been reported. The occurred accidents already in recent years lead to a probability of imperfection with superior values 1:500, when the socially acceptable one is of, in the maximum, 1:10000. Twelve spontaneous landslides already had been registered, twelve buildings had been demolished and about 110 building if they find interdicted for not offering conditions of security for habitation [22, 24].

In this type of construction, generally leaked ceramic blocks seated with the perforations in the horizontal line or blocks of concrete are used, with low

Fig. 1 Examples of
pathological manifestations
observed, namely, a detail of
masonry crushing between
foundation beam and ground
floor slab

compressive strength (2.5 MPa). The frequency of these accidents and the brusque
nature of the rupture, with gradual collapse, have generated fidget to the community
technique and, mainly, to the inhabitants of these constructions, that today live in
frightening for the uncertainty of the conditions of structural security of its
residences.

Rupture	Localization	Rupture cause
	Recife	Horizontal openings executed along the entire length of a central partition stop for the installation of conduits (1994)
	Jaboatão dos Guararapes	Loss of foundation block strength due to moisture expansion (1997)
	Olinda	Loss of resistance due to the degradation produced by the continuous action of sulphate ions on the concrete blocks (1999)
	Olinda	Rupture of the Foundation's ceramic blocks (1999) Seven fatalities occurred.
	Jaboatão dos Guararapes	Collapse of basements caused by the dismantling of running shoes due to the passage of sewage and rainwater (2011)
	Jaboatão dos Guararapes	There was a localized rupture of the basements in the region of the façade corresponding to the entrance of the building (2007)

Fig. 2 Examples of accidents occurred in the Metropolitan Region of Recife, Brazil

The established framework, which constitutes a serious social problem that afflicts many developing countries, demands the carrying out of consistent scientific research that allows a deep understanding of the problem and allows the creation of technical retrofitting interventions to avoid new accidents.

This constructive practice had an important impetus from the beginning of the 60's and its success was due to the lower cost compared to the construction with conventional reinforced concrete structure and the velocity of execution in the region at the time [23].

On the other hand, the search for cost minimization, the lack of quality control of the components and the construction procedures, together with the lack of specific design codes has been causing a series of pathologies and accidents over the last years.

With regard to retrofitting strategies, there is scarce information in the literature on the subject. The only reference of which authors have knowledge is the research developed at the School of Engineering of São Carlos (EESC-USP), which investigated the contribution of the coating in the strength of masonry prisms built with non-structural blocks [7, 21].

In local practice, what has been observed is the use of retrofitting solutions based on empirical knowledge that need more in-depth reflection on its effectiveness and applicability [4].

In this context, the work presents results of research developed within the framework of the FINEP/HABITARE Project entitled Development of Models for Retrofitting Masonry Buildings Constructed with Non-Structural Bricks. The Project was conducted by the Catholic University of Pernambuco—UNICAP—as executing agency, by the Technological Institute of Pernambuco—ITEP—as proponent, by the Secretariat of Science, Technology and Environment—SECTMA—as an intervener and the University of Pernambuco—UPE—and Federal University of Santa Catarina—UFSC—as co-executors.

2   Experimental Program

In order to evaluate the behaviour of resistant masonry elements used for structural purposes, blocks, prisms, wallets, walls and foundation elements experimentally tested. The prisms were made and tested in the Materials Laboratory of the Catholic University while the walls, walls and foundation elements were made and tested in the ITEP. The details of the elements and tests performed are presented below.

2.1   Units: Concrete and Ceramic Blocks

Ceramic and concrete blocks used in the research were of the same type as those usually employed in real resistant masonry buildings in the region. The dimensional

Table 1 Mean characteristics of the blocks testes

(a) Ceramic blocks—NBR 15270-1 [11]	
Length (mm)	190
Width (mm)	90
Height (mm)	190
Thickness of horizontal and vertical septa (mm)	7.0
Compression strength (MPa)	2.15
(b) Concrete blocks—NBR 6136 [12]	
Length (mm)	390
Width (mm)	90
Height (mm)	190
Thickness of horizontal and vertical septa (mm)	21.5
Thickness of the internal transverse septa (mm)	22.5
Thickness of external transverse septa (mm)	25.0
Compression strength (MPa)	2.30

characteristics of these blocks were obtained through tests of 60 ceramic blocks and 30 concrete blocks. Table 1 summarizes the results obtained for both type of blocks studied.

2.2   Fine Aggregate and Mortars

The sand used in the preparation of the mortars for laying and coating the tested models is usually found in the MRR and all the lot used in the development of the research was acquired from the same supplier. Table 2 summarizes the results of sand characterization.

The mortars used both in the laying of the blocks and in the coating, were defined from cement, lime and sand mixtures in proportions of 1:2:9, 1:1:6 and 1:0.5:4.5 by volume.

Table 3 presents the values of the compressive strength of the mortars, obtained through tests of 15 specimens in accordance with the Brazilian standards NBR 7215 [15], NBR 7222 [17]. The amount of cement used in the mortars was 220 and 150 kg/m^3 for the mixture proportions of 1:1:6 and 1:2:9, respectively.

Table 2 Characteristics of the natural sand used

Maximum characteristic size (mm)—NBR 7211 [14]	4.80
Fineness module—NBR NM 248 [19]	3.20
Unit mass (g/cm^3)—NBR NM 45 [20]	1.42
Specific mass (g/cm^3)—NBR 9776 [18]	2.60
Swelling—NBR 6467 [13]	1.25
Critical humidity (%)—NBR 6467 [13]	3.00
Powdery material content (%)—NBR 7219 [16]	1.26

Table 3 Characterization of mortars used

Item	Mean value		
	1:2:9	1:1:6	1:0.5:4.5
Compressive strength—MPa	4.00	5.80	6.23

In the case of concrete blocks, the laying grout cords were applied both to the longitudinal and to the transverse septa of the blocks, a situation that is usually referred as total settlement.

2.3 Steel Mesh and Connector

Two types of steel mesh were used as reinforcement of mortar coating in the prisms tested: one using galvanized steel and other with ribbed welded steel. The galvanized steel mesh is formed of wires with a diameter of 2.7 mm and a spacing of 5 cm in the horizontal direction and 10 cm in the vertical, making a steel area of 1.06 and 0.53 cm^2/m, respectively. The ribbed welded steel mesh had wires with a diameter of 4.2 mm and a spacing of 10 cm in the horizontal and vertical directions, making a steel section of 1.38 cm^2/m in both directions. Steel connectors were 5.0 mm in diameter.

2.4 Prisms

Approximately 500 prisms made with three blocks in vertically direction were tested—300 made with ceramic blocks and 200 with concrete blocks. The prisms were all capped at the top and bottom with cement paste in a thickness of 5 mm to obtain a uniform surface. Prisms were made in order to reproduce the conditions found in the daily practice of resistant masonry constructions in the region. The typology of prisms tested together with the corresponding acronym to identify each pattern in lab tests is presented below.

- Prisms of uncoated concrete and ceramic blocks (PSR);
- Prisms of concrete and ceramic blocks coated with 3.0 cm of mortar coating (PR30MM);
- Prisms of concrete and ceramic blocks coated with 3.0 cm of reinforced mortar (PCRTP—with ribbed welded steel—and PCRTG—with galvanized steel);
- Prisms of concrete and ceramic blocks coated with 3.0 cm of reinforced mortar with steel connectors (PCRTP-C and PCRTG-C);
- Prisms of concrete and ceramic blocks coated with 3.0 cm of mortar and an additional reinforced mortar layer with steel connectors (PRAATG–C e PRAATP–C).

All prisms were initially coated with a 5 mm layer of a scratch coating in a mixture proportion of 1:3 (cement and sand) and after 24 h they received an additional coating layer of 2.5 cm in a mixture proportion of 1:1:6 (cement, lime and sand) by volume. The prisms were submitted to a curing process under ambient conditions for a period of at least 28 days.

The execution of the coating of the prisms with a layer of 3.0 cm was carried out in four stages, as described below:

- Apply the scratch coating;
- Apply one layer of mortar (1.0 cm thick);
- Apply and install of steel meshes with connectors and
- Apply a second layer of 1.5 cm thick of mortar coating 1.5 cm leaving steel meshes fully immersed in the mortar.

The prisms that received a reinforced mortar layer over the existing unreinforced coating were initially made following the same procedure used in the prisms with mortar coating of 3.0 cm in thickness without steel meshes, which were coated in a single step by means of jigs wooden.

Once this step is completed and after a curing period of 28 days, transverse holes were made in the prisms through which steel connectors were inserted to install the steel meshes on the surface of the coating. Completed this operation, the second layer of coating mortar was applied over the steel meshes, leaving it fully involved and creating a final 6.0 cm thickness mortar layer. All prisms were capped with cement paste at the top and bottom. The transport of the specimens to the test machine required special care in order to avoid damages.

2.5 Wallets Specimens

The wallet is an element that better represent a real masonry wall because it contains all its parts, i.e.: bed and head joints and individual units lay in and bound together by mortar.

In order to analyse the influence of the mortar mixture proportions, the coating thickness and the reinforcement with steel meshes interlinked by connectors, 154 ceramic wallets were made. The specimens had dimensions of 0.09×0.60 1.20 m^3. Ceramic blocks of eight holes with dimensions of $9 \times 19 \times 19$ cm^3, with bed and head mortar joints made from a mixture proportion of 1:1:6 (cement, lime and sand) in volume. Figure 3 shows the types of tested wallets and Table 4 presents the characteristics of the wallets tested, all made with bed and head joints made with mortar of mixture proportion of 1:1:6 in volume. Fifteen specimens from each type were constructed.

The wallets were made using a three-course stage by day with bed and head mortar joints of 1.0 cm in thickness. They were built over steel 8 in.—H channel section, some of them filled of concrete to facilitate to apply mortar coating to the

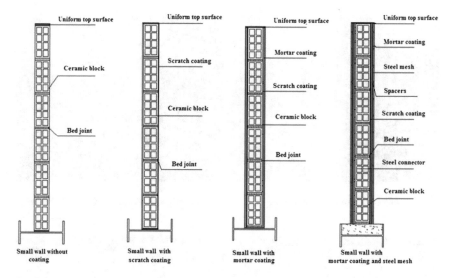

Fig. 3 Typical schemes of the tested walls

Table 4 Wallets characteristics

References	Scratch coating mixture proportion	Mortar coating mixture proportion	Mortar thickness on each side (cm)	Obs.
[1]	–	–	–	w/o coating
[2]	1:3	–	–	only w/scratch coating
[3]	1:3	1:2:9	1.5	w/o steel mesh
[4]	1:3	1:2:9	3.0	w/o steel mesh
[5]	1:3	1:1:6	3.0	w/o steel mesh
[6]	1:3	1:0.5:4.5	3.0	w/o steel mesh
[7]	1:3	1:1:6	3.0	w/o steel mesh w/additive
[8]	1:3	1:2:9	3.0	w/steel mesh
[9]	1:3	1:2:9	1.5	w/steel mesh
[12]	1:3	1:1:6	3.0	w/steel mesh

faces of the wall after the installation of the steel meshes of the same type of those used in the prisms. Figure 4 shows the steps of confection of the wallets tested and Fig. 5 exhibits the process of installation of steel meshes and connectors before the application of mortar coating layer, also showed in the figure [1].

Fig. 4 Wallets construction steps—elevation, scratch coating and mortar coating application

Fig. 5 Steps of drilling, installation of connectors, steel mesh, mortar coating and spacers

Special devices to transport the specimens and perform the works to smooth the top and bottom surface of the walls were created as it can be seen in Fig. 6.

The tests were carried out on a steel reaction frame with digitally controlled hydraulic loading system and digital data acquisition system. Vertical load was applied using hydraulic jacks with 200 mm maximum stroke and 1500 kN of compression capacity. The applying loading velocity used in the tests was 0.05 MPa/s.

3   Results and Discussion

The results from the performed tests are presented and discussed in the following sections.

Fig. 6 Details of the handling the specimen and execution of top bottom surface

3.1 Ceramic Brick Prisms

Figures 7, 8, 9 and 10 illustrate the rupture modes observed in the tested ceramic block prisms. As it can be observed, the ruptures are fragile and explosive, characterized by an immediate loss of the system's strength capacity soon after reaching the maximum load. For the coated prisms, it was observed that the cracking process starts in the horizontal septa of the blocks and from this moment, the two coating layers primarily carry the load. It should be noted, however, that several ways of rupture were observed, and it is a hard task to choose one that represents the universe of prisms tested. Several factors can influence the rupture process, such as: the quality of the workmanship used to construct the specimens, the thickness and uniformity of the bed mortar joints, among others. Nevertheless, the following ruptures modes can be highlighted as more frequent [10]:

- Coating cover rupture caused by excessive lateral displacement of bed mortar joints (Fig. 8, third photo);
- Rupture by detachment of the coating layers;
- Buckling rupture of mortar coating layers without connectors (Fig. 9, third photo).

Table 5 summarizes the rupture load of ceramic block prisms.

Fig. 7 Prisms of uncoated ceramic blocks: evolution of rupture

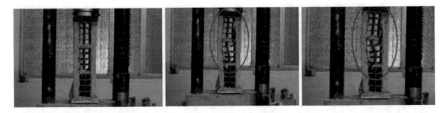

Fig. 8 Ceramic block prisms with coating: evolution of rupture

Fig. 9 Ceramic block prisms with coating and steel mesh

Fig. 10 Ceramic block prisms with reinforced mortar coating layer with connectors

As shown, the increase in load carrying capacity of the prism coating is significant. When the uncoated prism is compared to the uncoated prism, it is observed an increase of approximately 33% in its load capacity. When one compares the values obtained for the prism coated with that from the prism without any coating, there is an increase in the average of rupture load of approximately 130% that consubstantiates a significant increase in the load capacity generated by the mortar coating. The increment obtained with the installation of steel mesh inside the mortar coating was of 31 and 24%, for ribbed welded steel mesh and galvanized mesh, respectively.

Table 5 Rupture loads of ceramic block prisms—blocks with fbk = 2.15 MPa

Specimen	Average rupture load (kN)	Dispersion measures		Characteristic rupture load (kN)
		Standard deviation (kN)	COV (%)	
Uncoated prisms (PSR)	32.46	6.52	20.11	21.70
Prisms with scratch coating only (5 mm)	43.35	7.35	21.25	31.22
Prisms with 30 mm mortar coating (PR30MM)	74.93	13.90	18.55	52.00
Prisms coated with 3.0 cm of reinforced mortar with ribbed welded steel (PRCTP)	98.07	19.46	19.84	65.96
Prisms coated with 3.0 cm of reinforced mortar with galvanized steel (PRCTG)	92.86	21.59	23.25	57.23
Prisms coated with 3.0 cm of mortar and an additional reinforced mortar layer with ribbed welded steel and connectors (PRAATP–C)	205.44	16.06	7.82	178.93
Prisms coated with 3.0 cm of mortar and an additional reinforced mortar layer with galvanized steel and connectors (PRAATG–C)	150.18	25.33	16.86	108.39

The rupture observed in the prisms tested was always abrupt, for both coated and uncoated prisms. The rupture occurred due to the failure of the horizontal septa of the blocks by transverse tensile deformation followed by the collapse of the septa corresponding to the joints of mortar, causing the loss of equilibrium of the specimen. The coated prisms, reinforced with steel mesh and connectors also showed abrupt rupture, however with compressive loads well above the others. This is due to the presence of the connectors and their action of preventing the horizontal displacements of the septa.

The models with the highest average load of rupture, with lower coefficient of variation, were those reinforced with ribbed welded steel and connectors.

Analyses with varying thicknesses and mixture proportions of the mortar were additionally performed and the results obtained confirmed the above-mentioned observations. Further details can be found in Oliveira et al. [23].

3.2 Concrete Brick Prisms

Table 6 summarizes the rupture load of concrete block prisms. As shown, the increase in load capacity due to the prism coating was significant. In fact, in the case

Table 6 Rupture loads of concrete block prisms—blocks with fbk = 2.30 MPa

Specimen	Average rupture load (kN)	Dispersion measures		Characteristic rupture load (kN)
		Standard deviation (kN)	COV (%)	
Uncoated prisms (PSR)	86.22	11.47	13.30	67.30
Prisms with 30 mm mortar coating (PCR)	148.23	19.89	13.42	115.41
Prisms coated with 3.0 cm of reinforced mortar with ribbed welded steel (PRCTP)	223.43	17.07	7.64	195.26
Prisms coated with 3.0 cm of reinforced mortar with galvanized steel (PRCTG)	187.30	14.11	7.53	164.02
Prisms coated with 3.0 cm of reinforced mortar with ribbed welded steel and connectors (PCRTP-C)	251.70	36.43	14.47	191.59
Prisms coated with 3.0 cm of reinforced mortar with galvanized steel and connectors (PCRTG-C)	233.36	32.61	13.97	179.56
Prisms coated with 3.0 cm of mortar and additional reinforced mortar layer with ribbed welded steel and connectors (PRAATP–C)	371.36	38.37	10.33	308.06
Prisms coated with 3.0 cm of mortar and an additional reinforced mortar layer with galvanized steel and connectors (PRAATG–C)	392.42	34.80	8.87	335.03

coating without steel mesh, the increase in the average failure load, when compared to the uncoated prism, was approximately 72% whereas in the case of the coating with steel mesh inside the increase reached 159 and 117%, for ribbed welded and galvanized steel meshes, respectively. It was also possible to observe that the existence of the steel mesh inside the coating concurred to an increase in the load capacity of the coated prism. In the case of ribbed welded steel mesh, the average increase was approximately 51% and in the case of galvanized steel mesh, this increase was 27%. This behaviour indicates an important participation of the steel mesh that can be exploited for possible retrofitting works of real masonry building. Figures 11, 12, 13, 14 and 15 shown the rupture modes observed for the uncoated and coated concrete block prisms [2, 3].

In the case of uncoated concrete block prisms, the cracks observed were located on the faces of the blocks and on the settlement surface. The former presented a markedly random characteristic, while the latter presented a pattern with more regularity.

Fig. 11 Rupture of uncoated concrete block prisms

Fig. 12 Rupture of concrete block prisms with coating without steel mesh

Fig. 13 Rupture of concrete block prisms with coating and steel mesh

The rupture observed was less abrupt than that registered in the uncoated prisms made with ceramic blocks and in some of the prisms, displacements of the walls of the blocks were observed, as it can be seen in Fig. 9.

For the coated prisms, without addition of steel meshes inside the mortar coating, the most frequent rupture mode is indicated in Fig. 10. This type of rupture was characterized by a cracking process located on the front and back faces of the blocks, suggesting a rupture generated by transversal tensile strains.

For the prisms with steel meshes inside the mortar coating, the most frequent failure mode is shown in Fig. 13. It was similar to that observed for coated prisms

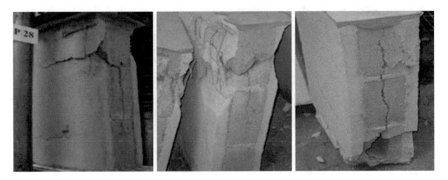

Fig. 14 Rupture of concrete block prisms with coating, steel mesh and connectors

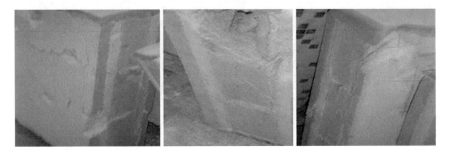

Fig. 15 Rupture of concrete block prisms coated with 3.0 cm of mortar and an additional reinforced mortar layer

without steel meshes. It deserves attention, however, the improvement in the performance of the prisms generated by the steel meshes inside the mortar coating and the fact that is was observed a slightly change in the failure mode that became less abrupt than that observed in prisms with unreinforced coating.

Another aspect that should be emphasized was the greater regularity in the value of the rupture load of the coated prisms with steel mesh that was not present in the prisms with coating but without steel mesh inside. This more regular behaviour can be attributed exclusively to the existence of steel meshes inside the mortar coatings. It is possible to infer that the coating with reinforced mortars should have contributed to the load capacity of the prisms since the beginning of the loading process, but their more effective participation occurred when the blocks were no longer able to withstand the applied load.

When the steel meshes were installed inside the mortar coating, it was observed an increase in the load capacity of the coated prism of approximately 27% in the case of a galvanized steel mesh and 51% in the case of ribbed welded steel mesh. This better performance of ribbed welded steel mesh can be attributed to its better adhesion conditions.

Table 7 Comparison between failure loads of concrete and ceramic block prisms

Comparison indicator (%)	Type of block	
	Concrete	Ceramic
Load increase after application of mortar coating	71.91	130.84
Load increase after the application of steel mesh without connectors	50.74	30.88
Load increase after the application of the connectors to the steel mesh	69.81	–
Load increase after the application of an additional reinforced mortar layer with connectors over existing mortar coating	164.76	174.18

The coating layer contributed significantly to the increase of the failure load of the prisms. The increment obtained with the installation of steel meshes inside the mortar coating also provided a significant improvement in the loading capacity of the studied prisms around 65%. Table 7 presents a comparison of the results of concrete block and ceramic block prisms.

Analysing data in Table 7, it is possible to observe the influence of the mortar coating on the ceramic block prisms was greater than on the concrete block one. The steel mesh provides more than 50% increase in failure load of concrete block prisms and a lower influence on the ceramic block prisms—30.88%. This increase in load capacity of ceramic block prisms can be attributed to the high strength of the ceramic when compared to the strength of the block itself. In the course of the tests, it was possible to observe that the ceramic blocks prisms collapsed before the prism reached the failure load and, consequently, the mortar coating was much more demanded. The lower contribution of the steel mesh to the increase in the load capacity of the ceramic block prisms can be explained by the fact that the steel mesh failed in buckling when the critical load was reached, taking, this way, the mortar coating layer together. The action of the connectors proved to be fundamental to combat the transverse tensile strains that motivate the rupture and the provision of some ductility to the specimen. For the concrete blocks prisms, the buckling of the steel meshes was partially counteracted by the connectors that interconnect the two steel meshes immersed in the mortar coating layers.

3.3 Wallet Specimens

Tables 8, 9 and 10 show the results of compressive tests in wallets specimens. The rupture mode of most of the wallet specimens occurred in the septum of the blocks in the upper region, close to the point of application of the load, and later they were accompanied by cracks in the interface of mortar coating and the scratch coating layers. Figure 16 shows details of the characteristic rupture mode of coated wallet specimens.

This behaviour is due to the tri-axial stress state to which the bed mortar joint is subjected as a consequence of its confinement between the blocks. This stress state

Table 8 Loads corresponding to first crack in the blocks of wallet specimens

Small walls	Average load of first crack (kN)	Dispersion measures		First crack characteristic load (kN)
		Standard deviation (kN)	COV (%)	
Uncoated specimen	41.40	11.60	27.90	22.30
Scratch coating specimen	67.90	19.60	28.80	35.60
Coating with mortar mixture proportion of 1:2:9 and thickness of 1.5 cm specimen	85.30	22.70	26.60	47.80
Coating with mortar mixture proportion of 1:2:9 and thickness of 3.0 cm specimen	96.10	34.50	35.90	39.20
Coating with mortar mixture proportion of 1:1:6 and thickness of 3.0 cm specimen	105.50	43.40	41.10	33.90
Coating with mortar mixture proportion of 1:0.5:4.5 and thickness of 3.0 cm specimen	170.30	54.20	31.80	80.90
Coating with mortar mixture proportion of 1:2:9 and thickness of 1.5 cm + reinforced mortar coating of 3.0 thickness layer with mixture proportion of 1:1:6 specimen	242.00	84.60	35.00	102.40
Coating with mortar mixture proportion of 1:2:9 and thickness of 3.0 cm + reinforced mortar coating of 3.0 thickness layer with mixture proportion of 1:1:6 specimen	254.80	57.50	22.60	159.90

generates horizontal tensile stresses to horizontal septa of the blocks, due to the mobilized its adhesion between with bed mortar joints. Thus, when tensile stress exceeds the tensile strength of these septa, they crack, distributing such stress to the others septa and to the mortar coating that tends to crack or detach, if there is no satisfactory adhesion, before the rupture of the specimen.

In the case of wallet specimens without reinforced mortar coating layer, cracks in the septa of the blocks occurred before cracks were observed in the mortar coating mortar, aspect that indicates an effective participation of the coating in the compressive behaviour of the specimen. For wallet specimens with reinforced mortar coating with connector, the initial crack occurred at the interface between mortar coating mortar and reinforcement mortar. This happened possibly due to the greater deformability of reinforced mortar layer and its positioning (without confinement) in relation to the core of the coated and confined specimens associated with the lower adhesion at the interface between the old and the new coating.

Taking into account the results of wallet specimens tested it is possible to formulate the following considerations.

Table 9 Loads corresponding to first crack in mortar coating of wallet specimens

Small walls	Average load of first crack (kN)	Dispersion measures		First crack characteristic load (kN)
		Standard deviation (kN)	COV (%)	
Uncoated specimen	–	–	–	–
Scratch coating specimen	–	–	–	–
Coating with mortar mixture proportion of 1:2:9 and thickness of 1.5 cm specimen	–	–	–	–
Coating with mortar mixture proportion of 1:2:9 and thickness of 3.0 cm specimen	123.70	34.00	27.50	67.60
Coating with mortar mixture proportion of 1:1:6 and thickness of 3.0 cm specimen	133.10	48.70	36.30	52.70
Coating with mortar mixture proportion of 1:0.5:4.5 and thickness of 3.0 cm specimen	240.80	42.60	17.70	170.50
Coating with mortar mixture proportion of 1:2:9 and thickness of 1.5 cm + reinforced mortar coating of 3.0 thickness layer with mixture proportion of 1:1:6 specimen	252.10	44.90	17.80	178.00
Coating with mortar mixture proportion of 1:2:9 and thickness of 3.0 cm + reinforced mortar coating of 3.0 thickness layer with mixture proportion of 1:1:6 specimen	206.60	59.70	28.80	108.10

3.4 Scratch Coating Influence

It was observed that the simple application of a scratch coating thin layer (5.0 mm of thickness) was capable to generate an average increase of 50.7% in the failure load of the wallet specimens, without, however, changing the sudden form of collapse. Figure 17 shows this behaviour.

By analysing the behaviour of the wallet specimens (Figs. 17 and 18), it was possible to observe an increase of the inclination of the load x displacement curve. The average stiffness of the wallet specimens without scratch coating layer was of 17.28 kN/m, while to the same specimens with scratch coating layer it was of 21.60 kN/m, representing an increase of 25%. In these figures, the stretches of the load-displacement curves located after the values of the maximum loads represent the unloading curve of the press and have no physical meaning.

Table 10 Failure load of wallet specimens

Small walls	Average failure load (kN)	Dispersion measures		Characteristic failure load (kN)
		Standard deviation (kN)	COV (%)	
Uncoated specimen	56.30	8.70	15.40	41.90
Scratch coating specimen	84.90	16.30	19.20	58.00
Coating with mortar mixture proportion of 1:2:9 and thickness of 1.5 cm specimen	130.40	24.40	18.70	90.10
Coating with mortar mixture proportion of 1:2:9 and thickness of 3.0 cm specimen	156.50	16.10	10.30	129.90
Coating with mortar mixture proportion of 1:1:6 and thickness of 3.0 cm specimen	168.30	33.30	19.80	113.40
Coating with mortar mixture proportion of 1:0.5:4.5 and thickness of 3.0 cm specimen	262.20	42.70	16.30	191.70
Coating with mortar mixture proportion of 1:2:9 and thickness of 1.5 cm + reinforced mortar coating of 3.0 thickness layer with mixture proportion of 1:1:6 specimen	321.00	47.70	14.90	242.30
Coating with mortar mixture proportion of 1:2:9 and thickness of 3.0 cm + reinforced mortar coating of 3.0 thickness layer with mixture proportion of 1:1:6 specimen	367.00	49.30	13.40	285.70
Coating with mortar mixture proportion of 1:1:6 and thickness of 3.0 cm + reinforced mortar coating of 3.0 thickness layer with mixture proportion of 1:1:6 specimen	417.09	62.99	15.10	313.35

The results show that the scratch coating layer generated a 50% increase in the load capacity of the wallet specimens and a 25% increase in its stiffness without, however, changing the abrupt failure mode.

3.5 Mortar Coating Mixture Proportion Influence

A comparison of the results of the wallet specimens according the mortar coating mixture proportion (1:2:9, 1:1:6 and 1:0.5:4.6) it was possible to observe a discrete increase in the load capacity of the specimen of about 7.5%, between the 1:2:9 and

Fig. 16 Rupture mode of coated wallet specimens

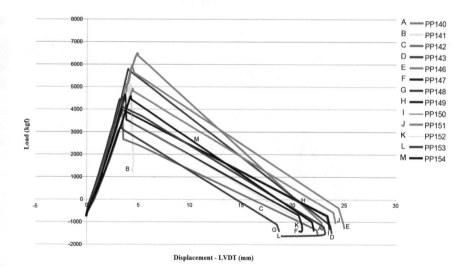

Fig. 17 Load versus displacement diagram of wallet specimens without scratch coating layer

1:1:6 mixture proportion and a considerable increase of 55.8%, between mortar mixture proportions of 1:0.5:4.5 and 1:1:6.

By analysing the compressive behaviour of the wallet specimen tested (Figs. 19, 20 and 21) it is possible to observe that there is an increase in the stiffness of the coated wallets with the increase of the cement content in the mortar mixture. The wallet coated with a mortar layer mixture proportion of 1:2:9 showed an average rigidity of 48 kN/m while those coated with 1:1:6 mortar mixture proportion showed an average rigidity of 58 kN/m—an increase of 21%. Wallet specimens

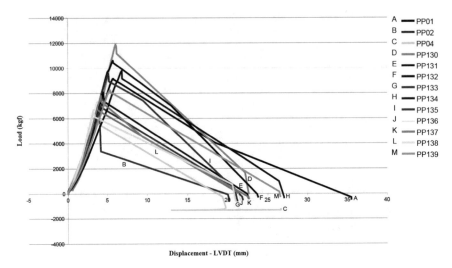

Fig. 18 Load versus displacement diagram of wallet specimens with scratch coating layer

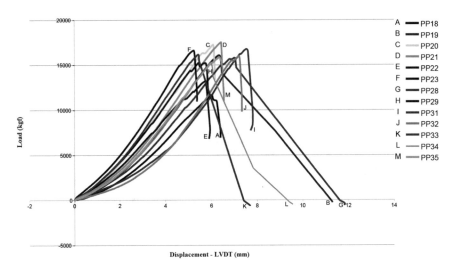

Fig. 19 Load versus displacement diagram of wallet specimens coated with a 3.0 cm in thickness mortar layer with a mixture proportion of 1:2:9

coated with mortar layer with mixture proportion of 1:0.5:4.5 presented average rigidity of the order of 64 kN/m, which represents an increase of 10% in relation to the previous one. In these figures, the stretches of the load-displacement curves located after the values of the maximum loads represent the unloading curve of the press and have no physical meaning.

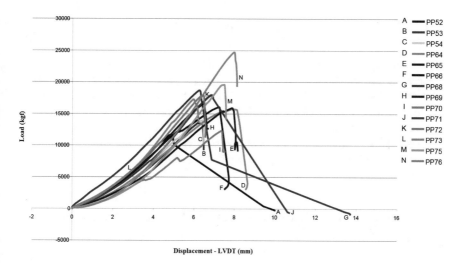

Fig. 20 Load versus displacement diagram of wallet specimens coated with a 3.0 cm in thickness mortar layer with a mixture proportion of 1:1:6

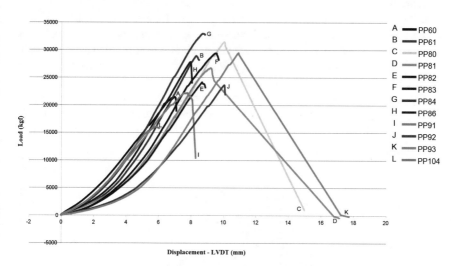

Fig. 21 Load versus displacement diagram of wallet specimens coated with a 3.0 cm in thickness mortar layer with a mixture proportion of 1:0.5:4.5

The increase in cement content of mortar coating layer mixture proportions generated increases in the load capacity and stiffness of wallet specimens, without, however, changing the abrupt failure mode.

3.6 Mortar Coating Layer Thickness Influence

By promoting a comparative analysis of the average strength of wallets as a function of the thickness of the mortar coating layer with mixture proportion of 1:2:9 with thicknesses of 1.5 and 3.0 cm, it can be observed that the failure load increased with the increment of the thickness. It can also be noted that this increase was of about 8.5%, between the thicknesses of 1.5–3.0 cm. The increases in failure loads of wallets, comparing their values with those from the model with only a thin scratch coating layer, were of 58 and 72%, respectively.

3.7 Influence of Reinforced Mortar Layer with Steel Meshes Inside

Wallets specimens with a 3.0 cm thick mortar coating layer in a mixture proportions of 1:2:9 and 1:1:6 were reinforced with steel meshes (10×10 cm^2) with 4.2 diameter wires interconnected through steel connector with 6.0 mm in diameter spaced of 20 cm and covered with a mortar layer with mixture proportion of 1:1:6.

The results show an expressive increase in the load capacity of the specimens, when reinforced with steel meshes interlocked by connectors. Increases of 134.50 and 147.82% were observed in the specimens with a mixture proportion mortar coating of 1:2:9 and 1:1:6, respectively.

Figures 22, 23, 24 and 25 show the typical force versus displacement curves of the various typologies of wallets tested. In these figures, the stretches of the

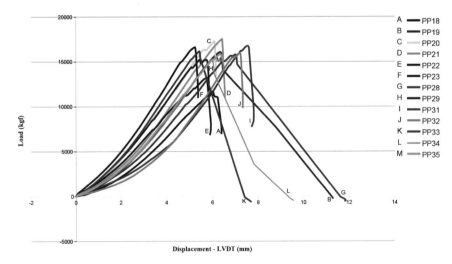

Fig. 22 Load versus displacement diagram of wallet specimens coated with a 3.0 cm in thickness mortar layer with a mixture proportion of 1:2:9 without reinforcement

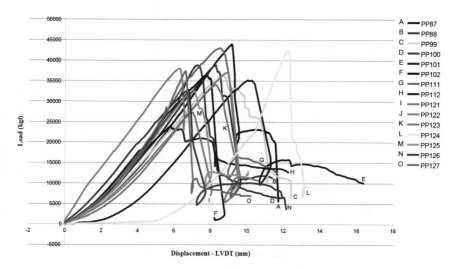

Fig. 23 Load versus displacement diagram of wallet specimens coated with a 3.0 cm in thickness mortar layer with a mixture proportion of 1:2:9 with reinforcement

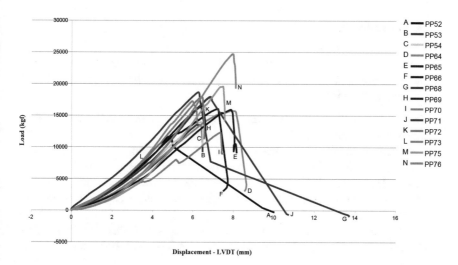

Fig. 24 Load versus displacement diagram of wallet specimens coated with a 3.0 cm in thickness mortar layer with a mixture proportion of 1:1:6 without reinforcement

load-displacement curves located after the values of the maximum loads represent the unloading curve of the press and have no physical meaning.

Observing the post-peak behaviour of the wallet specimen reinforced with steel meshes, one can note the importance of the connectors interlocking the steel meshes. While the 90° hooks of the steel connectors did not open, the specimens maintained the load capacity similar to that from specimens without reinforcement.

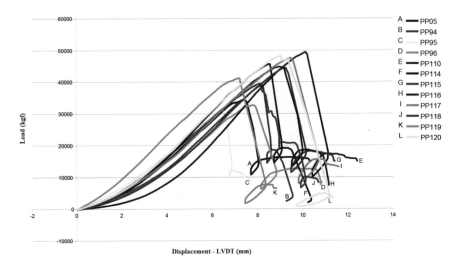

Fig. 25 Load versus displacement diagram of wallet specimens coated with a 3.0 cm in thickness mortar layer with a mixture proportion of 1:1:6 with reinforcement

4 Conclusions

Throughout the research, more than 500 prisms consisting of concrete and ceramic blocks, 154 ceramic blocks wallet specimens were tested. The tests performed allow the following considerations:

- The mortar coating layer contributed to increase of the load capacity of the tested models (prisms and wallet specimens);
- The incorporation of steel meshes inside mortar coating interlocked by steel connectors generated an additional increase of the load capacity of the elements tested. The connector has played a fundamental role in increasing the load capacity of the coated prisms and wallet specimens reinforced with steel meshes, without connectors; and
- Several forms of rupture of the tested prisms and wallet specimens were observed, and, this way, it could not be chosen a failure mode that representative those observed.

Without prejudice to the foregoing considerations, it is important that the following condition be observed:

- Resistant masonry (that where it is used non-structural blocks to carry load beyond its own weight) cannot be thought, in any situation, as a building process to support loads beyond its self-weight. Consistent with this assertion, all buildings constructed using this constructive technique must be retrofitting. Risk factors determined by whatever methodology should serve merely as indicators of the order or sequence in which these building must be retrofitted.

They, therefore, provide only a scale of priorities for the interventions to be carried out, which should be used with caution by the recovery process manager;

- The fact that the mortar coating layer contributes to increase the load capacity of a masonry wall serves merely to explain the reasons why the respective wall did not collapse yet. It does not authorize to certify the safety of the building that suffers from congenital failure, that is, to have been executed with non-structural blocks to carry loads;

- The use of the data and test results of the present research in specific solutions of recovery of buildings made with non-structural blocks is of the responsibility of the designer, and the researchers do not have any responsibility for this use.

Acknowledgements The authors are grateful to FINEP and CNPq for the financial support granted to perform the research.

References

1. Andrade, S.T.: Influence of coating characteristics on resistance to compression of masonry walls of ceramic sealing blocks. M.Sc. Thesis in Civil Engineering, University Federal of Pernambuco, Recife, Brazil (2007)
2. Araújo Neto, G.N.: Influence of coating mortar on resistance to axial compression of reinforced masonry prisms of concrete blocks. M.Sc. Thesis in Civil Engineering, University Catholic of Pernambuco, Recife, Brazil (2006)
3. Azevedo, A.A.C.: Comparative evaluation of the influence of the simple and armed coating on the compressive behaviour of prisms and wallets of ceramic sealing blocks. M.Sc. Thesis in Civil Engineering, University Catholic of Pernambuco, Recife, Brazil (2010)
4. Campos, M.D.: Solutions considerations for recovery of constructed buildings with resilient masonry. M.Sc. Thesis in Civil Engineering, University Federal of Pernambuco, Recife, Brazil (2006)
5. Ewing, B.D., Kowalsky, M.J.: Compressive behavior of unconfined and confined clay brick masonry. J. Struct. Eng.—ASCE **130**(4), 650–661 (2004)
6. Gumaste, K.S., Nanjunda Rao, K.S., Venkatarama Reddy, B.V., Jagadish, K.S.: Strength and elasticity of brick masonry prisms and wallettes under compression. Mater. Struct. **40**(2), 241–253 (2007)
7. Hanai, J.B., Oliveira, F.L.: Collapsing of ceramic block buildings. Téchne **115**, 58–63 (2006)
8. Kaushik, H.B., Rai D.C., Jain, S.K.: Stress–strain characteristics of clay brick masonry under uniaxial compression. J. Mater. Civ. Eng.—ASCE, **19**(9), 728–739 (2007a)
9. Kaushik, H.B., Rai D.C., Jain, S.K.: Uniaxial compressive stress–strain model for clay brick masonry. Curr. Sci. **92**(4), 497–501 (2007b)
10. Mota, J.M.V.: Influence of coating mortar on resistance to axial compression in ceramic block resistant masonry prisms. M.Sc. Thesis in Civil Engineering, University Federal of Pernambuco, Recife, Brazil (2006)
11. NBR 15270-1: Ceramic Components Part 1: Hollow Ceramic Blocks for Non-load Bearing Masonry—Terminology and Requirements, Rio de Janeiro, Brazil (2005)
12. NBR 6136: Plain Concrete Hollow Block for Reinforced Masonry—Specification, Rio de Janeiro, Brazil (2014)
13. NBR 6467: Aggregates—Determination of Swelling in Fine Aggregates—Method of Test, Rio de Janeiro, Brazil (2006a)

14. NBR 7211: Aggregates for Concrete—Specification, Rio de Janeiro, Brazil (2009)
15. NBR 7215: Portland Cement—Determination of Compressive Strength, Rio de Janeiro, Brazil (1997)
16. NBR 7219: Aggregates—Determination of Pulverulent Materials Content—Test Method. Rio de Janeiro, Brazil (1982)
17. NBR 7222: Mortar and Concrete—Determination of the Tension Strength by Diametrical Compression of Cylindrical Test Specimens, Rio de Janeiro, Brazil (2011c)
18. NBR 9776: Aggregates—Determination of Fine Aggregate Specific Gravity by Chapman Vessel—Method of Test, Rio de Janeiro, Brazil (1998)
19. NBR NM 248: Aggregates—Sieve Analysis of Fine and Coarse Aggregates, Rio de Janeiro, Brazil (2003)
20. NBR NM 45: Aggregates—Determination of the Unit Weight and Air-Void Contents, Rio de Janeiro, Brazil (2006b)
21. Oliveira F.L., Hanai, J.B.: Behaviour analysis of masonry walls constructed with ceramic sealing blocks. In: Proceedings of the International Seminar on Structural Masonry for Developing Countries, Belo Horizonte, Brazil (2002)
22. Oliveira, R.A., Silva F.A.N., Sobrinho, C.W.: Buildings constructed with resistant masonry in Pernambuco: current situation and future prospects. In: Monteiro B.S., Vitório J.A. P. (Organizers). SINAENCO-PE and Knowledge Production: Collection of Technical Articles, Recife, Brazil (2008)
23. Oliveira, R.A., Silva F.A.N., Sobrinho, C.W.: Resistant masonry: an experimental and numerical investigation of its compressive behaviour. In: Recife: FASA, Brazil (2011)
24. Oliveira R.A., Sobrinho, C.W.: Accidents with buildings constructed with resistant masonry in the Metropolitan Region of Recife. In: João Pessoa: DAMSTRUC, Recife, Brazil (2006)
25. Sumathi, A., Saravana Raja Mohan, K.: Study on the effect of compressive strength of brick masonry with admixed mortar. Int. J. ChemTech Res. 6(7), 3437–3450 (2014)

Ceramic Membranes: Theory and Engineering Applications

H. L. F. Magalhães, Antonio Gilson Barbosa de Lima, S. R. de Farias Neto, A. F. de Almeida, T. H. F. de Andrade and V. A. A. Brandão

Abstract Porous membranes are equipment used to separate different phases, restricting, totally or partially, the transport of one or more species present in a fluid solution. Separation processes can be classified in microfiltration, ultrafiltration, nanofiltration and reverse osmosis Filtration using porous membranes has presented promising results in many industrial sectors, especially in water treatment. This chapter provides theoretical and experimental information about ceramic and polymer membranes, with particular reference to separation process. Herein, several topics related to this theme, such as, theory, experiments, macroscopic mathematical modeling, and technological applications are presented and well discussed. CFD simulations of the water/oil separation process using a tubular ceramic membrane have been performed. The study clarified the importance of the CFD technique to elucidate the fluid flow phenomena in porous membrane as used in liquid filtration processes.

H. L. F. Magalhães · S. R. de Farias Neto · A. F. de Almeida
Department of Chemical Engineering, Federal University of Campina Grande,
Av. Aprígio Veloso, 882, Bodocongó, Campina Grande, PB 58429-900, Brazil
e-mail: hortencia.luma@gmail.com

S. R. de Farias Neto
e-mail: fariasn@deq.ufcg.edu.br

A. F. de Almeida
e-mail: arthur.filgueira@eq.ufcg.edu.br

A. G. Barbosa de Lima (✉) · V. A. A. Brandão
Department of Mechanical Engineering, Federal University of Campina Grande,
Av. Aprígio Veloso, 882, Bodocongó, Campina Grande, PB 58429-900, Brazil
e-mail: antonio.gilson@ufcg.edu.br

V. A. A. Brandão
e-mail: vanderson_agra@hotmail.com

T. H. F. de Andrade
Department of Petroleum Engineering, Federal University of Campina Grande,
Av. Aprígio Veloso, 882, Bodocongó, Campina Grande, PB 58429-900, Brazil
e-mail: tony.andrade@ufcg.edu.br

© Springer International Publishing AG 2018
J. M. P. Q. Delgado and A. G. Barbosa de Lima (eds.), *Transport Phenomena in Multiphase Systems*, Advanced Structured Materials 93,
https://doi.org/10.1007/978-3-319-91062-8_4

111

Keywords Membrane · Separation process · Filtration · CFD simulation

1 Foundations in Porous Membrane

Around the 1970s, in conjunction with the classical processes of separation, a new class of process, using synthetic membranes with selective barrier, emerged. Such membranes arise in an attempt to imitate existing natural membranes, with respect to the characteristics of selectivity and permeability [1].

Membranes are semipermeable and selective barriers, whose function is to separate phases, restricting, totally or partially, the transport of one or more species present in a solution.

The membranes are mainly developed from two distinct classes of materials: the organic, mostly polymers; and inorganic, such as metals and ceramics.

Separation processes using membranes are commonly driven by two forces: concentration gradient between two phases of a semipermeable membrane and pressure-driven in the membrane, which occur in microfiltration, ultrafiltration, nanofiltration and reverse osmosis processes [2].

Filtration, especially those using membranes as a filter mechanism, has presented good results due to the several characteristics that lead them to have the best cost/benefit ratio, simplicity of operation, low energy cost, long lifetime and uniformity of permeate quality throughout the process.

Park et al. [3] report that the efficiency of separation and economic viability in membrane processes depends on the cost, operating energy, permeate flux and lifetime of the membrane, which is directly related to the fouling, caused by the polarization effect by concentration, formation of a gel layer on its surface, adsorption and blocked pores.

In the literature, several studies have been reported on the development and application of membranes, for example, the experimental works of Ma et al. [4], Jiang et al. [5], Cheng et al. [6], Matos et al. [7] and Zhu et al. [8]. In addition, we can cite the following theoretical works that use computational fluid dynamics as a tool in the study of membrane separation processes, Darcovich et al. [9] and Geraldes et al. [10] using flat plate membranes; Serra et al. [11] and Serra and Wiesner [12] using circular membranes, and Bellhouse et al. [13], De Souza [14] and Magalhães [15], using tubular membranes.

1.1 Membrane Classification

Membranes can be classified into several categories assuming various configurations and modules depending on the characteristics of the substances to be used in the process. For each membrane category, there is a pre-established average pore diameter (Fig. 1). In microfiltration, the pore diameter range from 0.1 to 10 μm; for

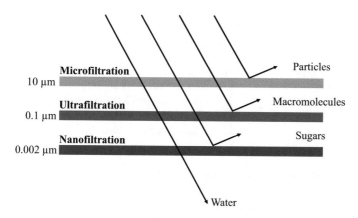

Fig. 1 Types of filtration according to the degree of particle retention

ultrafiltration, the pore diameter range from 0.001 to 0.1 µm; for nanofiltration membranes, the pore diameter range from 0.0005 to 0.002 µm; the choice depends on the morphology and size of the particles in the solution [16]. Following this topic will be detailed.

1.1.1 Nanofiltration Membranes

Nanofiltration consists in a membrane separation process indicated for molecules separation of a wide range of molecular weights, being used in several areas, from the fine chemical industry (separation of organic molecules) to high valued molecule recovery (drugs, enzymes and biocatalysts) [17]. Considered as a relatively new process, nanofiltration has characteristics that can be highlighted, such as its capacity of anion fractionation of low molecular weight with different sizes and valences through the ionic exclusion, and the fractionation of different organic components in aqueous solution [18].

1.1.2 Ultrafiltration Membranes

Ultrafiltration is a process indicated for solute particle size 10 times larger than the particle size of the solvent molecule and is intended for the concentration or fractionation of macromolecules [2]. In the water supply treatment, in which the organic contaminants is the main concern, the ultrafiltration process is considered adequate, presenting lower energy consumption, greater efficiency for the removal of organic pollutants and natural organic matter as a function of the molecular weight [19].

Table 1 Main characteristics of the separation process utilizing microfiltration membranes

Advantages	Disadvantages
Low cost; Low energy consumption; Possible integration to other treatment process; Capable of metal selective removal	Does not perform well for ionic substances and gases that permeate the system; Are subjected to chemical attacks; Effluent must present low contaminants concentration at the inlet of the membrane

1.1.3 Microfiltration Membranes

Microfiltration may be defined as a separation process that employs a permeable microporous membrane for phase separation as a barrier. It is a process applied to several industrial fields, which primarily uses ceramics in its fabrication because of its chemical stability, high durability and resistance to high temperature and pressure [20]. However, there are some advantages and disadvantages (Table 1) of the microfiltration process which must be taken into account according to the particle/process desired.

Application of microfiltration membranes in the most various processes and industrial fields is noticed in the literature, where the choice of membrane type and the material used in its preparation will depend on the characteristics of the substances to be treated [21–25].

1.1.4 Polymeric Membranes

Polymer membranes are characterized by their versatility in obtaining different morphologies in conjunction with low production cost, featuring numerous applications based on the characteristics of the material used for its manufacture. However, there are few commercial polymers that can be applied in the manufacture of membranes, thus, in several cases, modification or combination of polymers to obtain the adequate material to produce a membrane is performed [26].

In this sense, Leite et al. [27] report a growing use of second generation membranes produced from synthetic polymers (such as polyetherimide) presenting high chemical, thermal and chlorinated compounds resistance. In contrast, these membranes present low resistance to mechanical compaction.

Perles [28] highlights the use of polymer membranes in combustion cells, which are mechanisms responsible for converting Gibbs free energy variation of a redox reaction into electric energy, using the polymer membrane electrolytes as a tool. However, for polymer membranes to be well employed, the relationship between the molecular and morphological structure of the polymer must be well known for the well reactional development of the membrane.

In general, polymeric membranes, even showing some disadvantages in relation to the ceramic ones, hold numerous applications, such as, in dehydration process of ethanol, in fuel cells, in water treatment and in the food industry.

1.1.5 Ceramic Membranes

Ceramic membranes (Fig. 2) can be defined as a type of filter or a ceramic barrier separating two phases delimiting, totally or partially, the transport of one or several chemical species in the solution, where the separation capacity of the membranes will depend on porosity and selectivity [29].

Ceramic microfiltration membranes usually have pore sizes ranging from 0.2 to 0.8 μm and can be fabricated by any synthesizing particle method or sol-gel process, widely spread in the literature, presenting, as advantages, flexibility in the synthesis, uniformity in pore size and optimal reproducibility. The less diffused particle synthesis methods begin by coating a thin layer of particles to subsequently perform a high temperature treatment (1000–1600 °C) to partially synthesize the particles and create a porous separation layer [30].

Ceramic membranes present advantages over the polymeric ones, regarding their chemical inertia, biological stability and resistance to high temperatures and pressures. Also demonstrating advantages over the traditional methods of separation such as, distillation and centrifugation, showing low energy consumption, long lifetime, low space occupation and easy cleaning [31].

Zhu et al. [32] stress some advantages of ceramic membranes, when studying low cost composite membranes of microfiltration. The authors report that ceramic

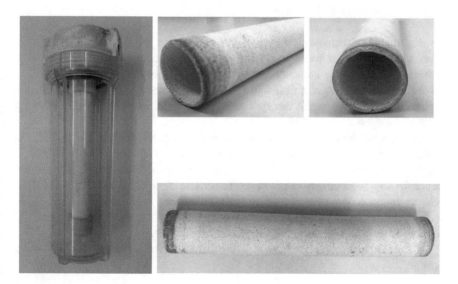

Fig. 2 Tubular ceramic membrane

membranes excel in the treatment of emulsions with particle diameters less than 20 µm, compared to other treatment techniques such as gravity sedimentation, adsorption and flotation. They also have several benefits compared to polymer membranes due to their mechanical performance and easy regeneration.

The pore shape, size and distribution are parameters that influence the permeability, in which the fragility is the main disadvantage of the ceramic membrane, which can be circumvented with a support material [33]. Further, Chen et al. [34] studying tubular ceramic membranes for nanofiltration, report that although they are particularly important for processes involving molecular separation under adverse conditions, they are usually made by the sol-gel process, which often presents problems of low efficiency and unsatisfactory control of the membrane properties. In their work, Chen et al. [34], present a strategy for confection of ceramic membrane, based on the atomic layer deposition and calcination, thus reducing this problem.

In synthesis, the ceramic membranes have high applicability, being commonly used in several separation processes, with good results.

1.2   Behind the Permeation Process

Separation processes using membranes as a filter involve various physical phenomena as permeation is initiated. Understanding some basic processes is fundamental.

1.2.1   Permeation and Diffusion

Hwang and Kammermeyer [35] report that permeation can be defined as the phenomenon that occurs when a particular species or component passes through another substance, occasionally occurring, but not necessarily by the diffusion mechanism. Since the term diffusion is specifically used to denote molecular diffusion, the term permeation emerges as a phenomenological definition involving various transport mechanisms.

Through the definition of the phenomenological permeability, it is possible to express schematically the profile of concentration through a membrane. Figure 3 graphically shows this behavior. In this figure, we can see the existence of the initial concentration (concentration of the contaminated fluid) and final concentration (concentration do fluid after treatment with membrane), respectively, Γ_1 and Γ_2, as well as the initial and final concentrations on the membrane surface, C_1 and C_2, with formation of a boundary layer near the membrane.

When the permeation mechanism is the diffusion, the permeate flux, F, can be written:

Fig. 3 Concentration profile through a membrane

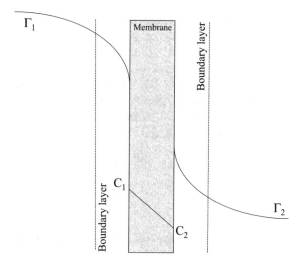

$$F = D_{AB}S\frac{C_1 - C_2}{e} \tag{1}$$

where S is the area and e is the membrane thickness.

Exist too, possible to study the resistance of the boundary layer separately by analyzing the membrane thickness on the Darcy permeability, K, defined by (Eq. 2):

$$R_m = \frac{e}{K} \tag{2}$$

where K is the Darcy permeability.

1.2.2 Polarization by Concentration

The effect of polarization by concentration is pertinent to any membrane separation process, being a phenomenon considered reversible. This effect occurs whenever a solution permeates a selective membrane, which provokes an increase in the solute concentration at the membrane interface, causing an increase in the solution osmotic pressure close to the membrane and reducing the driving force of separation [18].

According to Damak et al. [36], in a cross-flow filtration process with a membrane, the particles contained in the feed stream are convectively transported to the membrane surface, causing them to accumulate, until the equilibrium between the convective and diffusive flows is reached, forming a polarized layer by concentration. This becomes the major problem during filtration, because it causes the permeate flux to decline along the time.

Figure 4 illustrate schematically the appearance of the polarization layer by concentration. In this figure, we can see a feed flow at a given distance, δ, from the membrane surface, with a concentration, C_a. Is considered that, near the membrane, occurs the formation of a boundary layer that reaches its maximum value of concentration, C_m, being $J \cdot C$ the convective flux of solute in the membrane direction. If the solute is not completely retained by the membrane, there will be a diffusion flux return $D_{AB}(dC/dx)$, where the steady-state conditions will only be reached when the convective transport is equal to the sum of the permeate flux and diffusive transport [37].

The balance equation under permanent conditions can be written as follows:

$$J \cdot C - D_{AB} \frac{dC}{dx} = J \cdot C_p \qquad (3)$$

where C_p is the concentration after permeation and $J \cdot C_p$ is the flux of solute through the membrane.

The following boundary conditions, is considered:

$$x = 0 \Rightarrow C = C_m \qquad (4)$$

$$x = \delta \Rightarrow C = C_a \qquad (5)$$

Thus, integrating Eq. 3, we obtain:

$$\frac{C_m - C_p}{C_a - C_p} = \exp\left(\frac{Je}{D_{AB}}\right) \qquad (6)$$

where J is the volumetric flux and is the membrane thickness.

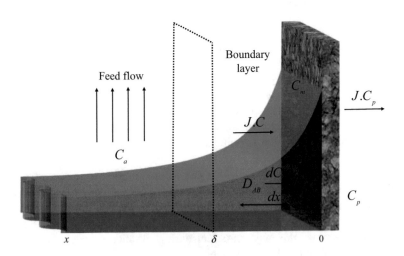

Fig. 4 Principle of polarization by concentration

This phenomenon of formation of the polarized layer depends mainly on the hydrodynamics of the system, the nature and size of the solute molecules and the membrane porosity and permeability; the latter limits the diffusive solute transport, which can cause irreversible negative effects such as: fouling by deposition, precipitation and bioinfustrations [37].

Adams and Barbano [38] studying microfiltration ceramic membrane, report that polarization by concentration can be attenuated by increasing the turbulence level in the flow.

1.2.3 Incrustation

Membrane operations have been widely used today; however, its large-scale and continuous use in filtration systems has suffered some limitations. The main problem is the membrane clogging, which provokes reduction of the filtrate flow and process efficiency as well as the increase in cost due to the energy consumption and materials for the cleaning process and backwashing.

There are three types of incrustation: deposition, precipitation and biofouling. When suspended solids deposition occurs, such as colloids, organic, corrosion products, iron hydroxide, algae and fine particulate matter, blockage of the feed channel of the membrane modules occur, often being difficult to remove, resulting in loss of performance.

Precipitation is a type of incrustation that occurs by precipitation of the soluble compounds present in the feed when its solubility limit is reached. Due to the concentration polarization, this effect intensifies near the surface of the membrane, being able to reach the limit of the salts solubility present in the solution causing precipitation; the most common are calcium carbonate, calcium sulfate, complexes of silica, barium sulphate, strontium sulfate and calcium phosphate.

Biofouling occurs due to the accumulation of organic material on the membrane surface, including cellular fragments, extracellular polymer substance and microorganisms, which result in the formation of biofilms, being equally detrimental to the membrane lifetime [39].

During membrane filtration, the formation of the polarized layer and incrustations result in declining the membrane performance with respect to quality and flow, reducing its reliability, since in applications with membranes, fouling control is essential [40].

1.3 Technological Applications of Ceramic Membranes

Due to its versatility, uniformity in permeate quality, chemical inertia, biological stability, high temperature resistance and long shelf life, ceramic membranes have the most varied technological applications, present in several industrial fields, such as the textile, paper and cellulose, oil, food and water treatment.

In pulp and paper industry, ceramic membranes are commonly used as the basis of advanced treatment processes because of their hydrophilic characteristics, high durability, high permeate fluxes, and thermal, chemical and mechanical stabilities. According to Ebrahimi et al. [41], paper processing produces substantial volumes of wastewater, requiring efficient recovery of waste, impurities and by-products prior to disposal into environment. Thus, membrane filtration is an effective alternative, which acts both in wastewater treatment and in the recovery of added value effluents.

Unlu et al. [42], in his work about the use of membranes for the reuse of waste water, report that in face of the various methods applicable to the dye wastewater treatment/recovery in textile effluents, membrane filtration technology is considered one of the most promising. Silva et al. [43] point out ceramic membranes as an important tool for the treatment of water from this sector. Meksi et al. [44] notice that textile wastewater has high staining, biochemical oxygen demand, chemical oxygen demand and salinity, which makes its treatment essential before its reuse or disposal in the environment, stressing out the membranes importance.

Lee et al. [45] report the use of ceramic membranes for water treatment. These authors argue that ceramic membranes are still less used in the treatment of wastewater to obtain potable water because of a historical conception about the high manufacturing cost of these membranes in comparison to the polymer membranes already consolidated in the market for this purpose. However, because the practice of sustainable development linked to the manufacture of ceramic membranes, it was possible to make membranes with better ratio cost-benefit, enabling the production of higher performance and more compact systems, especially in forms tubular, monolithic and in-plane-aligned, which currently present superior than performance the hollow fiber membranes and present a large permeation fluxes.

Kumar et al. [46] conducted a research on the treatment of milk wastewater using a low-cost tubular ceramic membrane. According to the authors, the dairy industry produces an enormous quantity of wastewater due to the different operations present in the unit that demand this resource, generating approximately 2.5 L of wastewater for each 1 L of processed milk. This residue contains a high concentration of nutrients, as well as high chemical and biological oxygen demand, as well as high total suspended solids. Therefore, it is necessary to carry out an appropriate treatment prior to disposal in the environment. In this context, membranes, especially the ceramic, appear as promising alternative technology, when compared to conventional treatment techniques, being a more economical process and requiring less space for operation. Further, we can cite its resistance to corrosive chemical products even in high temperatures.

In oil industry, membrane filtration technology is used in the processes of water treatment produced from oil and gas exploration and production activities. According to Jamaly et al. [47] oily wastewater is contaminated water at various concentrations of oil, and may contain fats, hydrocarbons and petroleum fractions, such as diesel oil, gasoline and kerosene, thus requiring adequate treatment for subsequent disposal. According to the specifications regulated by the Environmental Control Brazilian Agency, the content of oils and grease in effluents cannot exceed 20 mg/L [48]. Despite of the various technologies applied to the separation processes, the ceramic membranes have been distinguished by their high permeate index, acting as one of the final stages of this treatment process.

Damak et al. [36] point out the separation processes using membranes as one of the most significant developments in chemical and biological process engineering, evidencing their application in a wide range of industrial operations, in addition to those previously mentioned, such as: separation of organic molecules, pharmaceutical, biological macromolecules, colloids, ions and solvents.

2   Liquid Filtration with Membranes

2.1   Laminar Flow Through Porous Membrane

For laminar flow mechanisms in membranes, the hydrodynamic theory is considered simple. However, it is necessary to know information about distinct parameters, such as: average pore diameter, capillary mean length, tortuosity, pore size distribution, particle diameter, porosity, etc. Therefore, a phenomenological discussion of hydrodynamic flow through porous systems is necessary [35].

2.1.1   Darcy's Law

Considered as the law that describes the fluid flow through a porous system, Darcy's law demonstrates that the fluid flow rate is directly proportional to the pressure gradient, as follows:

$$\frac{V_J}{S \cdot t} = \frac{K}{\mu} \frac{\Delta P}{e} \tag{7}$$

where V_J is the volume, S represent the area, t is the time, μ viscosity, ΔP is the pressure drop, e is the membrane thickness and K correspond to Darcy's permeability.

Hwang and Kammermeyer [35] report that Darcy's law denotes that the flow resistance is due to viscous drag, and that the permeability parameter, K, contains all the properties of the porous system. Fluid viscosity is defined as the internal friction between fluid sheets flowing at different velocities, where for a fluid in laminar flow, such friction produces shear forces. When a fluid comes into contact with a solid surface, it adheres to the surface, resulting in a zero velocity, and as a consequence of the viscosity and fluid adhesion property, the solid surface experiences a drag force; the viscous resistance is, thus, a contrary force to the drag force.

Damak et al. [36, 49], Cunha [50], Pak et al. [51], De Souza [14], and others, used in their research, mathematical models whose the Navier–Stokes equations numerical solutions are connected to Darcy's law, written as a series resistance model.

2.1.2   Kozeny–Carman's Equation

By considering a porous membrane constituted by a capillary bundle with non-circular cross-section, that the fluid flow path well be tortuous, and using the concept of hydraulic radius, the following Darcy's law was derived:

$$\frac{V_J}{S \cdot t} = \frac{\varepsilon_p^3}{k'(1 - \varepsilon_p)^2 S_0^2 \mu} \frac{\Delta P}{e} \tag{8}$$

where S_0 is the surface area per particle volume and k' is a non-dimensional constant, dependent on the pore structure. When comparing the Eqs. (7) and (8), the Darcy's permeability, K, becomes:

$$K = \frac{\varepsilon_p^3}{k'(1 - \varepsilon_p)^2 S_0^2} \tag{9}$$

According to Hwang and Kammermeyer [35], modifications on the Kozeny's theory are numerous. The Kozeny–Carman equation is widely used for the case of laminar flow in macroscopic scale, especially in filtration problems.

2.1.3   Capillarity Model (Hagen–Poiseuille Equation)

When a porous membrane consists of straight cylindrical capillaries of equal size, the Hagen–Poiseuille equation can be applied directly to describe the fluid flow rate. It is defined as follows:

$$\frac{V_J}{t} = \frac{nS\pi r^4}{8\mu} \frac{\Delta P}{e} \tag{10}$$

where n is the number of capillary tubes per area unit, and r is the capillary tube radius.

2.1.4   Damak's Model

Occurring when the fluid moves in a well-defined trajectory, the laminar flow exhibits layers that individually preserve the characteristics of the fluid, in which the viscosity is responsible for weakening the tendency for turbulences to appear.

Thus, Damak et al. [36] report a model in which the permeation rate is expressed as a function of the pressure gradient, viscosity and flow resistance, as follows:

$$U_w = \frac{\Delta P}{\mu (R_m + R_p)} \tag{11}$$

where R_m is the membrane resistance. This parameter is a function of the membrane thickness and the porous media permeability. R_p is the resistance due to polarization by concentration defined, as follows:

$$R_p = r_p \delta_p \tag{12}$$

where δ_p is the concentration layer thickness and r_p is the resistance to the homogeneous concentration layers, determined by the Carmen–Kozeny equation (Eq. 8).

$$r_p = 180 \frac{(1 - \varepsilon_p)^2}{d_p^2 \varepsilon_p^3} \tag{13}$$

In the Eq. (13), d_p is the solute particles average diameter, and ε_p is the porosity of the polarization layer per concentration. Equation (13) is valid for disperse spherical particles (non-deformable) with porosity varying in the range $0.35 \leq \varepsilon_p \leq 0.75$.

Damak et al. [36], when studying a separation process using laminar tubular membrane, mentioned that the concentration polarization layer can be quantified by the empirical correlation for boundary layer thickness of concentration, δ_p, (Eq. 14), valid for Reynolds number varying between $300 < Re < 1000$ and Schmidt number ranging between $600 < Sc < 3200$.

$$\frac{\delta_p}{d_i} = 2 \left(\frac{z}{d_i} \right)^{0.33} (Re\,Sc)^{-0.33} Re_w^{-0.3} \left[\left(1 - 0.4377 \left(Sc^{-0.0018} Re_w^{-0.1551} \right) \right) \right] \tag{14}$$

where Re is the Reynolds number, Re_w represent the permeate Reynolds number, d_i is the internal diameter, Sc correspond to Schmidt number, and z is the axial coordinate.

The Reynolds permeation number is given by Eq. (15), as follows:

$$Re_w = \frac{\rho U_w D_{eq}}{\mu} \tag{15}$$

where the equivalent diameter can be given by:

$$D_{eq} = \left(\frac{\varphi}{1 - \varphi} \right) d_p \tag{16}$$

where φ is the membrane porosity, d_p is the average diameter of the particles that constitute the porous material, considering as spherical particles, and U_w is the permeation velocity.

By using the correlation of the boundary layer thickness it is possible to conclude that increasing the number of Reynolds leads to a reduction of the boundary layer thickness and an increase in the permeate flux, and that an increase in permeation Reynolds number and Schmidt number lead to increasing polarization by concentration [50].

2.2   Turbulent Flow Through Membranes

Early discussion is limited to laminar flow through porous media at low Reynolds number (typically $Re < 10$) for which a proportionality validity relationships exists between the fluid volumetric flow rate and pressure gradient.

Unlike the laminar flow regime, turbulent flow (flow at large Reynolds number) occurs when the fluid does not move along a well-defined path, producing momentum transfer. In this case, the deviation of the Darcy's law is clear.

Due to the existence of turbulent flows, the mass conservation and linear momentum equations only cannot adequately predict the oscillations resulting from this phenomenon. Therefore, it is necessary to consider other terms to the equation that predicts the fluid flow behavior. There are several models to predict turbulent flow, e.g., RNG k-ε model.

The RNG (*Renormalization Group Theory*) model emerged as an alternative of the k-ε standard model to flows with large Reynolds number, since various authors defend the idea that the k-ε turbulent model is inadequate to predict physical situations where there is rotational flux. For this condition, the model overestimates the kinetic energy dissipation, which results in a central recirculation region, smaller than that experimentally observed. Thus, the turbulent kinetic energy and kinetic energy dissipation rate equations (Eqs. 17 and 18) are defined by:

$$\frac{\partial}{\partial t}(\rho k) + \nabla \cdot \left(\rho \vec{U} k\right) = \nabla \cdot \left[\left(\mu + \frac{\mu_T}{\sigma_{kRNG}}\right)\nabla k\right] + P_k - \rho \varepsilon \tag{17}$$

and

$$\frac{\partial}{\partial t}(\rho \varepsilon) + \nabla \cdot \left(\rho \vec{U} \varepsilon\right) = \nabla \cdot \left[\left(\mu + \frac{\mu_T}{\sigma_{\varepsilon RNG}}\right)\nabla \varepsilon\right] + \frac{\varepsilon}{k}(C_{\varepsilon 1 RNG}P_k - C_{\varepsilon 2 RNG}\rho \varepsilon) \tag{18}$$

where k is the turbulent kinetic energy, ε is the turbulent kinetic dissipation rate, μ is the dynamic viscosity, ρ is the density and μ_T is the turbulent viscosity, given by:

Table 2 Typical values of the parameters in the RNG k-ε model

C_ε	C_μ	β	σ_k	σ_ε
1.9200	0.0850	0.0120	0.7179	0.7179

Source Ref. [52]

$$\mu_T = C_\mu \rho \frac{k^2}{\varepsilon} \tag{19}$$

where C_μ is an experimental calibration constant.

The coefficient $C_{\varepsilon 1 RNG}$ is given by:

$$C_{\varepsilon 1 RNG} = 1.42 - \frac{\eta\left(1 - \frac{\eta}{4.38}\right)}{\left(1 + \eta^3 \beta_{RNG}\right)} \tag{20}$$

where

$$\eta = \sqrt{\frac{P_k}{\rho \varepsilon C_{\mu RNG}}} \tag{21}$$

In Eq. (21), the parameter P_k is the production term by the shear effect given by:

$$P_k = \mu_T \nabla \vec{U} \cdot \left(\nabla \vec{U} + \nabla \vec{U}^T\right) + P_{kb} \tag{22}$$

where P_{Kb} is the fluctuation production term (Eq. 23) determined as follows:

$$P_{kb} = -\frac{\mu_T}{\rho \sigma_\rho} \vec{g} \cdot \nabla \rho \tag{23}$$

where $\vec{g}$ is the gravity acceleration vector, ρ the fluid specific mass, μ_T is the turbulent viscosity and σ_μ is the turbulent Prandtl number.

The values for the model constants, $C_{\varepsilon 2 RNG}$, $C_{\mu RNG}$, β_{RNG}, are specified on the Table 2.

3 CFD Applications: Separation Process by Porous Membranes

3.1 The Physical Problem

Herein, a theoretical study of a two-dimensional water/oil separation process using a tubular ceramic membrane is performed. Figure 5 illustrate the physical problem to be considered. It consists in a ceramic tubular membrane with 0.03 m diameter, and 3 m length, according to the dimensions used in the work of Damak et al. [53], being, therefore, subjected to a tangential flow of oily water. As the membrane has angular symmetry, only one cross section was used in the YZ plane.

For the numerical simulation, was generated a computational domain 2D, using ICEM CFD software, as shown in Fig. 6. The mesh was created with a total of having a total of 205,056 elements and 180,000 nodes, with concentration of the elements near the surface of the membrane. Figure 7 shows the boundaries conditions used in the modeling: inlet, concentrate outlet and permeate outlet.

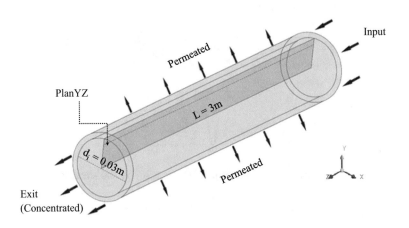

Fig. 5 Geometric representation: tubular membrane 2D

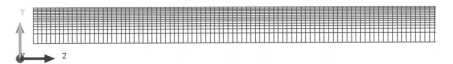

Fig. 6 Two-dimensional mesh of the cross-section of the tubular membrane

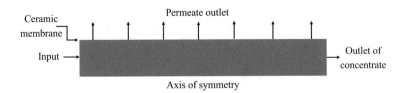

Fig. 7 Representation of membrane boundaries

3.2 *Mathematical Modeling*

For this purpose, the following considerations were assumed:

- Viscosity and density of the fluids are constant and equal to those of the pure solvent;
- The diffusion coefficient is considered constant;
- Laminar and permanent flow;
- Incompressible and isotropic porous system;
- The local permeation velocity is determined according to the in series resistance theory;
- Concentration layer is assumed to be homogenous and the Carmen–Kozeny expression is valid;
- Gravitational effect is neglected;
- Axial velocity profile is parabolic, according to laminar and fully developed flow at the tubular membrane inlet cross section;
- Symmetry in relation to the angular coordinate;
- The porous medium obstruction by the solute is neglected;
- Solute adsorption on the membrane contact surface along with possible reactions are neglected;
- The resistance resulting from concentration polarization, which is due to the concentration layer formed at the membrane-fluid interface was considered;
- Wall roughness was disregarded.

3.2.1 Governing Equations

From the considerations, the mass and linear momentum conservation equations can be written as:

$$\nabla \cdot \left(\rho \vec{U} \right) = 0 \tag{24}$$

where ρ and $\vec{U}$ are, respectively, fluid density and velocity vector.

- Momentum equation

$$\nabla \cdot \left(\rho \vec{U} \otimes \vec{U} \right) - \nabla \cdot \left(\mu \nabla \vec{U} \right) = -\nabla p + \nabla \cdot \left[\mu \left(\nabla \vec{U} \right)^{T} \right] \tag{25}$$

where p is pressure and μ is the solution viscosity.

- Energy conservation equation

$$\frac{\partial}{\partial t}(\rho H) + \nabla \cdot \left(\rho \vec{U} \, H\right) - \nabla \cdot (\Gamma_e \nabla T) = S^H \tag{26}$$

where Γ_e is the effective thermal diffusivity, H is enthalpy and S_H is the heat source term. The first term of Eq. (26) is responsible by the energy accumulation; the second relates to heat transfer by convection and the third one is responsible for heat transfer by diffusion.

- Mass transfer equation

$$\vec{U} \cdot \nabla C = D_{AB} \nabla^2 C \tag{27}$$

where C, is the solution (oil) concentration, D_{AB} is the mass diffusivity, considered constant for each Schmidt number (Sc), as follows:

$$D_{AB} = \frac{\mu}{Sc\rho} \tag{28}$$

3.2.2 Boundary Conditions

(a) Inlet conditions (mixture, z = 0)

Fully developed flow is assumed at the tubular membrane inlet section. Thus, radial velocity is null and axial velocity is given as follows:

$$U_z(0, y) = 2\overline{U}_z \left[1 - \left(\frac{y}{R}\right)^2\right] \tag{29}$$

where y is the radial coordinate, R is the tubular membrane internal radius and $\overline{U}_z$ is the average velocity, determined by:

$$\overline{U}_z = \frac{Re\,\mu}{\rho R} \tag{30}$$

where Re is Reynolds number.

Initial oil concentration in the mixture, C_0, will be given as follows:

$$C = C_0 \tag{31}$$

and inlet temperature; will be as follows:

$$T = T_0 \tag{32}$$

(b) Outlet conditions (Concentrated, z = L)

In the outlet section, we consider:

$$P = P_0 \tag{33}$$

$$\frac{\partial C}{\partial z} = 0 \tag{34}$$

$$\frac{\partial U_z}{\partial y} = 0 \tag{35}$$

$$\nabla T = 0 \tag{36}$$

(c) Symmetry condition (y = 0)

The following symmetry conditions were assumed at the central axis of the tube:

$$\frac{\partial U_z}{\partial y} = 0 \tag{37}$$

$$\frac{\partial C}{\partial y} = 0 \tag{38}$$

$$U_y = 0 \tag{39}$$

$$\frac{\partial T}{\partial y} = 0 \tag{40}$$

(d) Transverse plane

Due to the presented symmetry condition by the membrane, it was admitted the following conditions for the transverse planes:

$$\frac{\partial U_z}{\partial x} = 0 \tag{41}$$

$$\frac{\partial C}{\partial x} = 0 \tag{42}$$

$$\frac{\partial T}{\partial x} = 0 \tag{43}$$

(e) Porous wall tube (y = R)

In the porous wall, zero axial velocity was assumed (non-slip condition) and a temperature gradient null. Then, we can write:

$$U_z = 0 \tag{44}$$

$$\nabla T = 0 \tag{45}$$

Through the porous wall, radial velocity U_y is equal to the permeation velocity U_w, as follows:

$$U_y = U_w = \frac{\Delta P}{\mu (R_m + R_p)} \tag{46}$$

where R_m and R_p are, respectively, the membrane resistance and the resistance from concentration polarization, and ΔP is the transmembrane pressure.

The equation responsible for the mass transport (Eq. 47) [54] was included in the model as a source term. This, term is given by:

$$U_w R_r C = D_{AB} \frac{\partial C}{\partial y} \tag{47}$$

where R_r is the solute intrinsic retention by the membrane that can be assumed to be constant for a membrane-solute system.

The transmembrane pressure (Eq. 48) is defined as being the difference between the average permeate pressure, $\overline{P}_p$, and the external pressure to the membrane P_{ex}, as follows:

$$\Delta P = \overline{P}_p - P_{ex} \tag{48}$$

The membrane resistance, R_m, the resistance from the polarization by concentration, R_p, the concentration layer thickness, δ_p, and the resistance, r_p, determined when the concentration layer is considered homogenous, are given by Eqs. (2), (12), (13) and (14).

Damak et al. [49] used Eq. 14 to determine local variation of the boundary layer thickness by polarization, δ_p, so that the equilibrium between the convective and the diffusive flows was reached when $[(C - C_0)/C_0] < 0.001$.

According to Damak et al. [49], the parameters (Re, Sc, Re_w and z/d) from Eq. (14) correspond to a membrane separation system for liquid ultrafiltration, low particle concentration and laminar flow in the wall of the porous tube.

	Parameters	
Fluid	Density, ρ (kg/m^3)	997
	Viscosity, μ (cP)	50
	Molecular weight, MM (kg/kmol)	18.02
Membrane	Porosity, ε_p (–)	0.35
	Permeability, K (m^2)	3.33×10^{-11}
	Thickness, e (m)	0.01

Table 3 Fluid and membrane parameters used in the simulations

Permeation Reynolds number	R_{ew}	0.1
Initial concentration (kg m^{-3})	C_0	1
Temperature (°C)	T_0	25
Intrinsic solute retention by the membrane	R_r	1
Oil droplet average diameter	d_p	51
Schmidt number	Sc	1000
Atmospheric pressure (atm)	P_0	1

Table 4 Parameters used in the simulations

Case	Re (–)
01	1000
02	600
03	300

Table 5 Conditions used in the two-dimensional simulation with tubular membrane

3.2.3 Fluid and Membrane Data and Simulated Cases

Important parameters used in the study of the 2D flow correspond to the thermal and physico-chemical properties of the fluid and the membrane. The parameters are given in Table 3.

In the simulations carried out, (Ansys CFX 15 software), the permeation Reynolds number, the average diameter of the oil droplet, d_p, Schmidt number, initial concentration, permeability, temperature, and intrinsic solute retention by the membrane, R_r, were kept constant according to Table 4, varying the axial Reynolds number, Re, as shown in Table 5. All the cases were simulated in isothermal conditions.

3.2.4 Results Analysis

(a) Concentration profile

In Fig. 8, the concentration profiles in different axial positions are represented, z/L (0.25, 0.50, 0.75 and 1.00), where L = 100d (tubular membrane length). It is

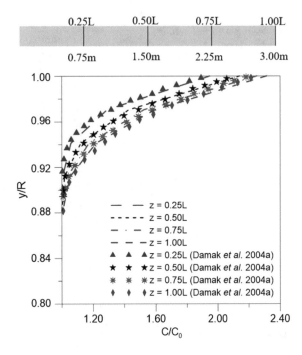

Fig. 8 Concentration profile inside the membrane at different axial positions (case 1)

verified that the concentration increases with axial position and that the behavior of the simulated results is similar to that one reported in the literature.

In Fig. 9, concentration profiles for different Reynolds numbers on the axial position z = 0.50L are illustrated. It is expected that as the permeation velocity increases a larger number of particles will be conducted convectively for near the membrane surface. Further, according to the resistance model, the permeate flux increases with transmembrane pressure. Therefore, higher Reynolds number provokes an increase in the transmembrane pressure, and consequently, concentration in the membrane surface also will increase. From this figure, we can see that results agreed well with those obtained in the literature.

(b) **Thickness of the concentration boundary layer**

Figure 10 illustrate the behavior of the thickness of the concentration boundary layer as a function of the non-dimensional distance z/d, for Reynolds numbers 300, 600 and 1000. It is observed that the boundary layer grows with increasing the non-dimensional distance z/d, and its thickness drops when Reynolds number increases. This behavior provokes an increase in the transmembrane pressure, and thus, we have an increase in the transport by convection of a higher number of particles to near oh the membrane surface, as observed by Damak et al. [36]. A reduction of 0.035 in the non-dimensional boundary layer thickness, δ_p/d, was observed when the Reynolds number ranged from 300 to 1000.

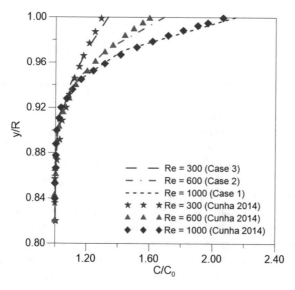

Fig. 9 Reynolds number effect on the concentration profile inside the membrane ($z = 0.50L$) (cases 1, 2 and 3)

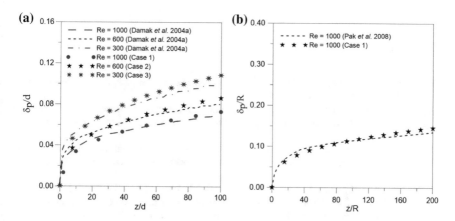

Fig. 10 Thickness of the concentration boundary layer versus the non-dimensional variable z/d for different Reynolds numbers

4   Final Considerations

Ceramic membranes are considered in the literature as one of the most promising technologies in the present day. They are always present mainly in the stages of separation and effluents treatment from the most varied industrial sectors. They can be produced in order to assume various configurations and modules, adapting continuously the processes needs. Membrane ceramic present diverse characteristics such as chemical inertia, biological stability, manufacturing flexibility, high permeability index and high temperature and temperature resistance, which provides high applicability, this versatility combined to cost-benefit ratio makes this equipment very competitive in comparison to many others already consolidated in the market.

In general, the topics presented in this chapter discuss the universe of application in which the ceramic membranes are inserted. The growing applications concern with sustainable industrial progress being the driving force for the continuous development of new research and technologies in this field of study. Besides, information about different models applied to filtration process in porous membranes are discussed, and CFD applications are related.

From the simulated results, we can conclude that the mathematical model was able to predict the behavior of the fluid in the water/oil separation process through the tubular membrane, and that, concentration increases with the axial position and with increasing Reynolds number. The increase in the Reynolds number leads to a reduction in the thickness of the concentration boundary layer.

Acknowledgements The authors are grateful for financial support provided by CNPq, CAPES and FINEP (Brazilian Research Agencies). We also acknowledge scientific support from the authors mentioned along this chapter.

References

1. Habert, A.C., Borges, C.P., Nobrega, R.: Membrane separation processes. E-papers, Brazil, Rio de Janeiro (2006) (in Portuguese)
2. Timoteo, J.F.J.: Anodizing for the production of ceramic membranes, p. 69. Masters dissertation, Mechanical engineering, UFRN, Rio Grande do Norte, Brazil (2007) (in Portuguese)
3. Park, J.Y., Jin Choi, S., Reum Park, B.: Effect of N2-back-flushing in multichannels ceramic microfiltration system for paper wastewater treatment. Desalination **202**(1–3), 207–214 (2007)
4. Ma, W., Guo, Z., Zhao, J., Yu, Q., Wang, F., Han, J., Pan, H., Yao, J., Zhang, Q., Samal, S. K., De Smedt, S.C., Huang, C.: Polyimide/cellulose acetate core/shell electrospun fibrous membranes for oil-water separation. Sep. Purif. Technol. **177**, 71–85 (2017)
5. Jiang, Y., Hou, J., Xu, J., Shan, B.: Switchable oil/water separation with efficient and robust Janus nanofiber membranes. Carbon **115**, 477–485 (2017)
6. Cheng, Q., Ye, D., Chang, C., Zhang, L.: Facile fabrication of superhydrophilic membranes consisted of fibrous tunicate cellulose nanocrystals for highly efficient oil/water separation. J. Membr. Sci. **525**, 1–8 (2017)
7. Matos, M., Gutiérrez, G., Lobo, A., Coca, J., Pazos, C., Benito, J.M.: Surfactant effect on the ultrafiltration of oil-in-water emulsions using ceramic membranes. J. Membr. Sci. **520**, 749–759 (2016)
8. Zhu, X., Dudchenko, A., Gu, X., Jassby, D.: Surfactant-stabilized oil separation from water using ultrafiltration and nanofiltration. J. Membr. Sci. **529**, 159–169 (2017)
9. Darcovich, K., Dal-Cin, M.M., Ballevre, S., Wavelet, J.P.: CFD assisted thin channel membrane characterization module. J. Membr. Sci. **124**(2), 181–193 (1997)

10. Geraldes, V., Semião, V., Pinho, M.N.: Numerical modeling of mass transfer in slits with semi-permeable membrane walls. Eng. Comput. **17**(3), 192–217 (2000)
11. Serra, C.A., Wiesner, M.R., Laîné, J.M.: Rotating membrane disk filters: design evaluation using computational fluid dynamics. Chem. Eng. J. **72**(1), 1–17 (1999)
12. Serra, C.A., Wiesner, M.R.: A comparison of rotating and stationary membrane disk filters using computational fluid dynamics. J. Membr. Sci. **165**(1), 19–29 (2000)
13. Bellhouse, B.J., Costigan, G., Abhinava, K., Merry, A.: The performance of helical screw-thread inserts in tubular membranes. Sep. Purif. Technol. **22–23**, 89–113 (2001)
14. De Souza, J.S.: Theoretical study of the microfiltration process in ceramic membranes, p. 134. Doctoral Thesis, Process Engineering, UFCG, Paraíba, Brazil (2014) (in Portuguese)
15. Magalhães, H.L.F.: Study of the thermofluidodynamics of the treatment of effluents using ceramic membranes: modeling and simulation, p. 102. Masters dissertation in Mechanical Engineering, Federal University of Campina Grande, Campina Grande, Brazil (2017) (in Portuguese)
16. Taketa, T.B., Ferreira, M.Z., Gomes, M.C.S., Curvelo, N.: Production of biodiesel by ethyl transesterification of vegetable oils and their separation and purification by ceramic membranes. In: VIII Brazilian Congress of Chemical Engineering in Scientific Initiation, Uberlândia, Minas Gerais, Brasil (2009) (in Portuguese)
17. De Carvalho, R.B., Cristiano, P.B., Nobrega, R.: Formation of double-spreading cellulosic flat membranes for nanofiltration and reverse osmosis processes. Polymers **11**(2), 65–75 (2001) (in Portuguese)
18. Lopes, A.C.: Degradation study of polymeric membranes of commercial nanofiltration by sodium hypochlorite, p. 92. Master dissertation, Federal University of Rio de Janeiro, Rio de Janeiro, Brazil (2006) (in Portuguese)
19. Mierzwa, J.C., Da Silva, M.C.C., Rodrigues, L.D.B., Hespanhol, I.: Water treatment for public supply by ultrafiltration: comparative evaluation through the direct costs of implantation and operation with the conventional and conventional systems with activated carbon. Sanitary Environ. Eng. **13**(1), 78–87 (2008) (in Portuguese)
20. Rosa, D.S., Salvini, V.R., Pandolfelli, V.C.: Processing and evaluation of the properties of porous ceramic tubes for microfiltration of emulsions. Ceramics **52**(322), 167–171 (2006) (in Portuguese)
21. Makabe, R., Akamatsu, K., Nakao, S.: Classification and diafiltration of polydispersed particles using cross-flow microfiltration under high flow rate. J. Membr. Sci. **523**, 8–14 (2017)
22. Yang, X., Zhou, S., Li, M., Wang, R., Zhao, Y.: Purification of cellulase fermentation broth via low cost ceramic microfiltration membranes with nanofibers-like attapulgite separation layers. Sep. Purif. Technol. **175**, 435–442 (2017)
23. Wang, X., Wang, C., Tang, C.Y., Hu, T., Li, X., Ren, Y.: Development of a novel anaerobic membrane bioreactor simultaneously integrating microfiltration and forward osmosis membranes for low strength wastewater treatment. J. Membr. Sci. **527**, 1–7 (2017)
24. Suresh, K., Pugazhenthi, G., Uppaluri, R.: Fly ash based ceramic microfiltration membranes for oil-water emulsion treatment: parametric optimization using response surface methodology. J. Water Process Eng. **13**, 27–43 (2016)
25. Suresh, K., Pugazhenthi, G.: Cross flow microfiltration of oil-water emulsions using clay based ceramic membrane support and TiO_2 composite membrane. Egypt. J. Petrol. **14**(1), 1–10 (2016)
26. Becker, C.M.: Obtaining and characterizing sulphonated polyelectrolytes based on styrenic copolymers for polymer membranes, p. 96. Masters dissertation, Federal University of Rio Grande do Sul, Porto Alegre, Brazil (2007) (in Portuguese)
27. Leite, A.M.D., Ito, E.N., Araújo, E.M., Lira, H. De L., Barbosa, R.: Obtaining microporous membranes from nanocomposites of polyamide 6/national clay. Part 1: influence of clay presence on membrane morphology. Polymers **19**(4), 271–277 (2009) (in Portuguese)

28. Perles, C.E.: Physical and chemical properties related to the development of Nafion® membranes for PEMFC fuel cell applications. Polymers **18**(4), 281–288 (2008) (in Portuguese)
29. Silva, A.A., Melo, K.S., Maia, J.B.N.: Study of the water/oil separation potential of alumina tubular ceramic membranes through analysis of flow and turbidity measurements. In: 2° Brazilian Congress of P & D in Oil & Gas. Rio de Janeiro: UFRJ, Rio de Janeiro, Brazil (2003) (in Portuguese)
30. Cui, J., Zhang, X., Liu, H., Liu, S., Yeung, K.L.: Preparation and application of zeolite/ceramic microfiltration membranes for treatment of oil contaminated water. J. Membr. Sci. **325**(1), 420–426 (2008)
31. Silva, F.A., Lira, H.L.: Preparation and characterization of cordierite ceramic membranes. Ceramics **52**(324), 276–282 (2006) (in Portuguese)
32. Zhu, L., Chen, M., Dong, Y., Tang, C.Y., Huang, A., Li, L.: A low-cost mullite-titania composite ceramic hollow fiber microfiltration membrane for highly efficient separation of oil-in water emulsion. Water Res. **90**, 277–285 (2016)
33. Maia, D.F.: Development of ceramic membranes for oil/water separation, p. 111. Doctorate Thesis in Process Engineering, Federal University of Campina Grande, Campina Grande, Brazil (2006) (in Portuguese)
34. Chen, H., Jia, X., Wei, M., Wang, Y.: Ceramic tubular nanofiltration membranes with tunable performances by atomic layer deposition and calcination. J. Membr. Sci. **528**, 95–102 (2017)
35. Hwang, S.T., Kammermeyer, K.: Membranes in Separation. Wiley, Canada (1975)
36. Damak, K., Ayadi, A., Schmitz, P., Zeghmati, B.: Modeling of cross-flow membrane separation processes under laminar flow conditions in tubular membrane. Desalination **168**, 231–239 (2004)
37. Mulder, M.: Basic Principles of Membrane Technology, 1st edn. Kluwer Academic Publishers, Netherlands (1996)
38. Adams, M.C., Barbano, D.M.: Effect of ceramic membrane channel diameter on limiting retentate protein concentration during skim milk microfiltration. J. Dairy Sci. **99**(1), 167–182 (2016)
39. Oliveira, D.R.: Pre-treatment of the reverse osmosis process using microfiltration and investigation of membrane cleaning and recovery techniques, p. 129. Master dissertation, Federal University of Rio de Janeiro, Rio de Janeiro, Brazil (2007) (in Portuguese)
40. Ahmed, S., Seraji, M.T., Jahedi, J., Hashib, M.A.: Application of CFD for simulation of a baffled tubular membrane. Chem. Eng. Res. Des. **90**(5), 600–608 (2012)
41. Ebrahimi, M., Busse, N., Kerker, S., Schmitz, O., Hilpert, M., Czermak, P.: Treatment of the bleaching effluent from sulfite pulp production by ceramic membrane filtration. MDPI Membr. **7**, 1–15 (2015)
42. Unlu, M., Yukseler, H., Yetis, U.: Indigo dyeing wastewater reclamation by membrane-based filtration and coagulation processes. Desalination **240**, 178–185 (2009)
43. Silva, M.C., Oliveira, R.C., Lira, H.L., Freitas, N.L.: Obtaining a ceramic membrane to treat effluent from the textile industry. Electron. J. Mater. Process. **9**, 81–85 (2014) (in Portuguese)
44. Meksi, N., Ben Ticha, M., Kechida, M., Mhenni, M.F.: Using of ecofriendly α-hydroxycarbonyls as reducing agents to replace sodium dithionite in indigo dyeing processes. J. Clean. Prod. **24**, 149–158 (2012)
45. Lee, M., Wu, Z., Li, K.: Advances in membrane technologies for water treatment: materials, processes and applications. In: Advances in Ceramic Membranes for Water Treatment, vol. 1, pp. 43–82, 1st edn. Woodhead Publishing, England (2015)
46. Kumar, R. V., Goswami, L., Pakshirajanb, K., Pugazhenthi, G.: Dairy wastewater treatment using a novel low cost tubular ceramic membrane and membrane fouling mechanism using pore blocking models. J. Water Process Eng. **13**, 168–175 (2016)
47. Jamaly, S., Giwa, A., Hasan, S.W.: Recent improvements in oily wastewater treatment: progress, challenges, and future opportunities. J. Environ. Sci. **37**, 15–30 (2015)
48. CONAMA N°20/ART.21, RE. Standard CONAMA n°20, de June 18, Brazil (1986) (in Portuguese)

49. Damak, K., Ayadi, A., Zeghmati, B., Schmitz, P.: Concentration polarisation in tubular membranes—a numerical approach. Desalination **171**(2), 139–153 (2004)
50. Cunha, A.L.: Treatment of effluents from the petroleum industry via ceramic membranes—modeling and simulation, p. 201. Doctorate Thesis in Process Engineering, Federal University of Campina Grande, Campina Grande, Brazil (2014) (in Portuguese)
51. Pak, A., Mohammad, T., Hosseinalipour, S.M., Allahdinib, V.: CFD modeling of porous membranes. Desalination **222**(1–3), 482–488 (2008)
52. Ansys: CFX 15, Solver Theory Guide. Ansys, Japan (2015)
53. Damak, K., Ayadi, A., Zeghmati, B., Schmitz, P.: New Navier-Stokes and Darcy's law combined model for fluid flow in cross-flow filtration tubular membranes. Desalination **161**(1), 67–77 (2004)
54. Minnikanti, V.S., Dasgupta, S., De, S.: Prediction of mass transfer coefficient with suction for turbulent flow in cross flow ultrafiltration. J. Membr. Sci. **154**(2), 227–239 (1999)

Unsteady State Heat Transfer in Packed-Bed Elliptic Cylindrical Reactor: Theory, Advanced Modeling and Applications

R. M. da Silva, Antonio Gilson Barbosa de Lima, A. S. Pereira, M. C. N. Machado and R. S. Santos

Abstract Transport phenomena through porous media has been of continuing interest for scientists, researchers and engineers due to the wide range of industrial applications. This chapter presents information about unsteady-state heat transfer and fluid flow inside of packed-bed reactors. The topics covered are related to fundamentals of porous media, chemical reactors, including mathematical modeling and applications. Emphasis is placed on packed-bed elliptic-cylindrical reactor. Based on the concept of local thermal equilibrium, a general mathematical formulation for a pseudo-homogeneous heat transfer model written in elliptic-cylindrical coordinates along with the numerical solution of the governing equation has been presented. Herein, an overview of currently-used models and the pertinent transient conductive transport processes inside the reactor were explored. A numerical example of heat transfer and fluid flow in a multiphase system (elliptic-cylindrical reactor filled with particles) was performed, and results of the

R. M. da Silva
Federal Institute of Education, Science and Technology of Paraíba,
Access Highway PB 426, Rural area, Princesa Isabel, PB 58755-000, Brazil
e-mail: rodrigo.silva@ifpb.edu.br

A. G. Barbosa de Lima (✉)
Department of Mechanical Engineering, Federal University of Campina Grande,
Av. Aprígio Veloso, 882, Bodocongó, Campina Grande, PB 58429-900, Brazil
e-mail: antonio.gilson@ufcg.edu.br

A. S. Pereira
Baiano Federal Institute of Education, Science and Technology, Highway PB 420,
Rural area, Santa Inês, BA 45320-000, Brazil
e-mail: antonildo.pereira@ifbaiano.edu.br

M. C. N. Machado
Department of Chemical, State University of Paraiba, R. Das Baraúnas, 351,
Campina Grande, PB 58429-500, Brazil
e-mail: ceicamachado3@gmail.com

R. S. Santos
Rural Federal University of the Semi-Arid, Av. Francisco Mota, 572,
Mossoró, RN 59625-900, Brazil
e-mail: rosilda.santos@ufersa.edu.br

© Springer International Publishing AG 2018
J. M. P. Q. Delgado and A. G. Barbosa de Lima (eds.), *Transport Phenomena in Multiphase Systems*, Advanced Structured Materials 93,
https://doi.org/10.1007/978-3-319-91062-8_5

temperature distribution inside the equipment at different instant of process are presented and discussed.

Keywords Reactor · Porous media · Packed-bed · Elliptic-cylindrical Finite volume · Unsteady-state

1 What Is a Porous Media?

1.1 Foundations

Material is considered porous medium if [1]:

- Contains relatively small voids (pores) inside a solid or semi- solid matrix. The pores usually contain one fluid or a mixture of various fluids.
- It is permeable to some types of fluids, i.e. these fluids must penetrate the porous medium through one side and emerge on the other side. Porous media with this property are called permeable porous media. The level of ease for a fluid flows through a particular porous medium depends on the characteristics of the fluid and medium, and the thermodynamic conditions in which fluid flow occurs.

Depending on the type of interconnections between the elements (particles) that constitute a particular porous medium, it is possible classify it as follows [2]:

(a) **Unconsolidated porous medium**: It is a porous medium in which is possible distinguish visually its constituent elements and physically separate them. This type of material can be created by depositing particles within a container, as shown in Fig. 1a. The term normally designated for this type of porous medium is filled or stuffed bed, which refers to the manner in which it is generated. The continuous phase is associated with the voids of the porous medium. The structure is defined by the relative positioning of the particles that forms the discontinuous solid phase.

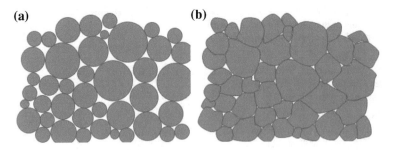

Fig. 1 a Initial state and **b** final state of a particulate filler after sintering by compression and/or heating

(b) **Consolidated porous medium**: It is a porous medium in which the solid phase is continuous or it is not possible to distinguish the particles that form the structure of the medium. It can be formed in different ways, for example, by dissolving a solid matrix, or by compressing and/or heating the particles of a non-consolidated medium (Fig. 1b).

In general, porous media can be found in nature and many industrial applications. Virtually all solid materials can be considered as porous, except for metallic structures, dense rocks and some plastics [1]. In certain scientific and technological practices, it is of fundamental importance to have suitable models that can predict the behavior of the porous media and the transport phenomena that occur inside of them. Some examples of areas where this knowledge is advantageous are:

(a) Chemical process engineering: fixed bed reactors, filtration, drying, fluidized bed reactors, chromatography, absorption/adsorption, ion exchanges, fuel cells, catalytic converters to reduce pollutant emissions in automobiles, absorption columns with and without chemical reaction, distillation columns with or without chemical reaction;
(b) Environmental engineering: migration of contaminants in soil and through groundwater, irrigation, cleaning of soils by steam injection, incineration;
(c) Natural resources: production of natural gas and oil, runoff in water mines.
(d) Mechanical engineering: thermal insulation, combustion involving pyrolysis of reactive or non-reactive materials, tribology and lubrication, nuclear reactors with gas cooling, solidification or melting of binary mixtures, dehumidification, sintering and aggregation of particles by compression and heating;
(e) Civil construction: penetration of moisture in porous materials and development of protection strategies to avoid the degradation of these materials by diffusion of water through of them, analysis of water retention of dams and fluid flow in them.

With a deeper understanding of transport phenomena occurring within a porous environment, it is possible to implement more appropriate changes or to develop more desirable processes for a sustainable development based on the lower emission of pollutants in the environment and higher energy efficiency and rational consumption of raw materials.

1.2 Important Parameters in Porous Media

The processes involving porous media are influenced decisively by the structure of these means. Thus, any model to adequately describe a porous medium must consider this aspect, however more complex become the models. Thus, it is often necessary to adopt simplifications in the representation of the structure of the porous medium taking care not to lose essential characteristics.

Usually to describe a porous medium in terms of modeling it is necessary to use some parameters. The importance of these parameters is to physically delineate the type of porous medium that is intended to be studied, in order to communicate to the mathematical model its real configuration. In this sense, some parameters are described below.

1.2.1 Porosity

This parameter is one of the most important in the characterization of a porous medium on a macroscopic scale. It is defined as the ratio between of the volume of voids, V_v, and the total volume of the sample, V_T, as follows:

$$\varepsilon = \frac{V_V}{V_T} \tag{1}$$

Specifically, the effective porosity is defined as the volumetric fraction associated with the channels through which the fluid flow occurs, which in practice refers to the interconnected channels. In unconsolidated media, generally, the total and effective porosities are considered approximately equal, already for the consolidated porous media, these two parameters may be quite different, depending on how the medium was generated. For example, there are consolidated media in which exist commonly closed or isolated pores. Due to this fact, the effective porosity is very different from the real one.

1.2.2 Permeability

When the movement of the fluid depends on the medium, we are talking about the resistance that the medium offers to the fluid flow due to the structure of this medium. This condition is characterized by a parameter, called permeability, which is independent of the fluid properties. The permeability property implies that the characteristic dimensions of the void spaces traversed by the fluid are much larger than the dimensions of the molecules of the fluid passing therethrough. The structure of porous media depends strongly on their origin.

1.2.3 Specific Surface Area

The specific surface area, A_{sp}, is defined as the ratio of the surface area of void space to the total volume of porous sample. It represents a measure of the contact area between the solid phase and the fluid phase, which is a very important aspect in a large number of processes. For example, in industrial adsorption, this parameter is a measure of adsorption capacity and strongly influences the performance of certain processes, such as fixed bed reactors with catalyst particles or chemical absorption units, usually characterized by this parameter associated with a of mass transfer coefficient between the phases [1].

1.2.4 Sphericity

The geometric characteristics of the particles constituting a porous medium, namely geometric shape and distributions, strongly influence the behavior of the non-consolidated porous media.

The equivalent diameter, D_p, which is defined as the diameter of the sphere with the same volume of particle [3] can be determined, among other forms, by the ratio $D_p = 6/A_{sp}$ if the specific surface area is known. An equivalent nominal diameter may also be obtained by sieving if the measurement of an equivalent diameter is not available. This expression for the diameter is more used for particles with geometry closer to the spherical, however, for particles with more irregular shape, we can use sphericity factor φ defined by the following expression:

$$\varphi = \frac{V_P}{D_P S_P} \qquad (2)$$

where values of this factor differ for particles of different forms. In the Eq. (2), V_p is the medium particles total volume, D_P is the equivalent diameter and S_p represents aggregate surface area of particles.

Since in many applications, the particles forming a packed-bed do not have a uniform diameter, but a size distribution, it is common practice to determine an average value for the particle diameter of the filling bed, which will function as its equivalent diameter. The reflection of this approximation is that there will be, for modeling facilitation purposes, a system in which the actual bed with distribution of various particle sizes will be represented by one system with particles containing an average diameter.

2 Chemical Reactors

Chemical reactor is the brain of most chemical processes. Their design and operation usually determine the success or failure of the whole process. In a process, in general, the raw materials are delivered to the chemical reactor in specified condition of temperature, pressure, and concentration of species. The chemical reactor is an essential component in which the lower value-added feedstock is converted into products of high commercial value.

The reactor is usually followed by separation equipment, with the task of separating the products from the non-reactive raw materials and byproducts from the reactor. Modern chemical processing plants has highly integrated operation, as they are used to manufacture chemicals and pharmaceuticals products, oil refining, or microelectronic devices. Although a reactor to achieve efficiency in production, its design and operation are also sometimes influenced by conflicting energy consumption minimization goals as well as quantity of raw material or product that must be kept in storage. In general, reactors can be classified in terms of design, type of reaction and operation. Here are some important aspects of reactors are discussed.

2.1 Batch Reactors, Continuous Agitation Tank and Plug-Flow Type

In general, chemical reactors can be modeled using three main reactor archetypes: batch, continuous agitation tank and plug flow. These three types of reactors are defined under certain idealized assumptions regarding fluid flow. The batch reactor and the continuous stirring tank are ideally assumed to be well mixed, meaning that the temperature, pressure and specific concentrations are independent of the spatial position within the reactor. In turn, the plug flow reactor describes a special type of flow, in which the fluid is well mixed in the radial direction and varies only in the direction of the axial length or direction of the length of the tube.

The reactors, because of its design complexity, for practical purposes, can always be framed in one of these three types cited previously, which are considered simpler for design purposes. The material and energy balance of these three reactors are systems of first order equations, nonlinear ordinary differential equations, nonlinear algebraic equations, or in some situations, differential algebraic equations. This large simplification, which is usually standard in initial design, is intended to disregard the momentum balance and to give a more careful treatment of the flow profile inside the reactor. Concentration, temperature and pressure are the usual dependent variables that are solved as a function of time or distance as the independent variable along the reactor.

Sometimes the reactor model is quite simple, such as describing a single reaction in an isothermal reactor. In these cases, and a single ordinary differential equation or algebraic equation describes the concentration of a single species or the extent of a single reaction. Often the design involves many reactions and non-isothermal operation, and systems of coupled ordinary differential equations or algebraic equations are needed to describe the temperature, pressure and concentrations of the species. Regardless of complexity, the design problem is approached in the same way. There are many cases, except for the simplest ones, where numerical methods are needed to solve the governing equations. Fortunately, high-level programming languages as well as commercial codes are readily available and can easily be used to solve complex models.

2.2 Homogeneous and Heterogeneous Reactions

Reactors are classified also with respect to the number of phases in which the reaction occurs. In homogeneous reactions, the reagents and products are in a single phase. Examples include: steam fission of ethane in ethylene, and photochemical reactions of chlorocarbons in the troposphere leading to destruction of ozone by gas phase reactions; enzymatic isomerization of glucose to fructose, and esterification of an acid and alcohol to produce an ester by liquid phase reactions; the melting of limestone and charcoal to produce calcium carbon, and the interfacial reaction between strontium oxide and silicon dioxide to form strontium silicate by solid reactions.

In turn, heterogeneous catalytic reactions such as zeolite-catalyzed cracking of high-boiling crude oil and iron-catalyzed ammonia synthesis from hydrogen and nitrogen, involve two phases. Sometimes three phases are present simultaneously during the reaction, for example gas synthesis ($CO + H_2$) reactions over iron-based catalysts that are suspended in a high molecular weight alkane with the intent to control the reaction temperature in moderate values and dissolve the reaction product. Multiphase reactions need not involve heterogeneous catalysts. For example, the reaction of liquid p-xylene and gaseous O_2 to produce liquid terephthalic acid occurs through free radical intermediates in the liquid phase inside the reactor.

2.3 Batch, Semi-batch and Continuous Operation

As for the operation of the reactors, they can be classified as batch, semi-batch and continuous operations. In batch operation, the reactor is loaded with reagents, the reaction occurs, and after some processing time the reactor contents are removed as product. The batch reactors, shown in Fig. 2, are often used for liquid phase reactions, and the manufacture of small quantity of high value-added products, such as special fine chemicals, pharmaceuticals and fermentation products. Batch reactors are also used in situations where it is impractical to implement continuous

Fig. 2 Representative diagram of batch reactor

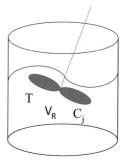

processes. The great flexibility of batch processing allows the reactor to be used for manufacturing of different products [3].

The batch cycle begins by charging the reactants into the vessel and often warming the reactants to the reaction temperature. The cycle usually ends by bringing the contents to the discharge temperature, emptying the container and cleaning it before the next loading of reagents. The product can be formed or not during these preparation steps. Further they generally involve manual labor, so that the manufacturing costs can be considerably higher than the that corresponding for a continuous process.

The semi-batch process is similar to the batch process, except for the addition of feed that occurs during the batch cycle. The products can also be removed during the semi-batch process. The addition/removal control allows controlling the rate of reaction or release of heat during the reaction. The semi-batch reactor can also provide a more complete use of reactor volume in reactions such as polymerizations that convert low density reagents into higher density products during the course of the reaction.

The continuous stirring tank reactor (CSTR) is shown in Fig. 3 [3]. In this type of reactor, the reagents and products flow into the reactor continuously, and the reactor contents are assumed to be well mixed. Consideration of the well mixed condition can be performed more easily for liquids than for gases, so CSTR's are often used for liquid phase reactions. The fluid composition and the temperature undergo change when passing from the feed stream into the interior of the reactor; the composition and temperature of the effluent stream are identical to those of the reactor.

CSTR is used extensively in situations where intense agitation is required, such as the addition of a gaseous reagent to a liquid by transfer between the bubbles and the continuous liquid, and the suspension of a solid or a second liquid within a continuous liquid phase. Polymerization reactions are sometimes performed in CSTR's. It is common to employ cascade or series CSTRs where the effluent from the first reactor is used as feed for the second and so on in the cascade (Fig. 4). The cascade allows a high conversion of reagent, while minimizing the volume of the reactor. In the Figs. 2, 3 and 4, C_j is the concentration of component j, Q is the volumetric flowrate, T is the temperature, V_R is the reactor volume, the subscript f represents the feed, and the numerical subscripts represents process stages.

Fig. 3 Schematic diagram of a CSTR

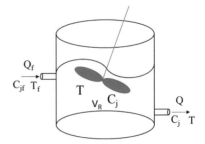

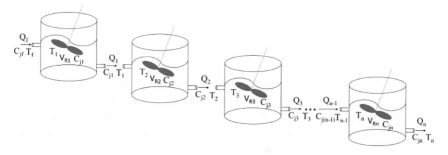

Fig. 4 Schematic diagram of continuous agitation tank reactors in series

The plug flow reactor (PFR) is a constant cross-section tubular reactor as shown in Fig. 5. Under turbulent flow conditions, the velocity profile becomes plug type, which greatly simplifies the energy and mass balance. The velocity, composition and temperature are functions only of the axial position, and the steady-state and dynamic profiles along the equipment can be studied.

The plug-flow type reactors are used for gas phase and liquid phase reactions. If the PFR is filled with a porous catalyst with fluid flow into the void space being turbulent, the reactor is referred to as a fixed bed-reactor. An isothermal plug-flow type reactor generally leads to greater conversion of reagent per unit volume than that of the continuous stirring tank, and therefore there is a more efficient use of the volume in a PFR. For this reason, PFR's are used in situations that require high capacity and high conversion rate, and in cases involving high exothermic or endothermic reactions. In bundled pipes of small diameter, the PFR can be placed in a furnace for endothermic reactions, or surrounded by a high-temperature boiling fluid for exothermic reactions. PFR's are the preferred choice reactors for gas-phase reactions due to gas mixing problems in a CSTR [4].

Fig. 5 Schematic diagram of a plug-flow type reactor

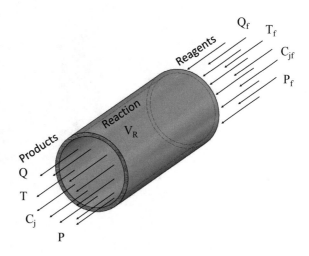

3   Heat Transfer in Fixed Bed Reactor

3.1   Basic Concepts

Fixed bed reactors can be treated as being equipment in which a fluid flows percolating a particulate solid medium (porous medium). The aim is to promote heat and mass transfer between the phases, in order to obtain products with interest of the chemistry industry. The porous medium in this type of reactor is stationary, differentiating, basically of a fluidized bed in which the particles are suspended during the contact between the phases. The fixed bed reactors are very present in catalytic processes involving gaseous reagents [5].

In simplified terms, a fixed bed reactor can be described as a cylindrical tube filled in a stationary and compact form by catalyst particles with a gaseous reagent flowing through this porous medium to promote fluid-particle interactions [6]. It is an unconsolidated porous medium.

Due to wide range of scale of applications of this equipment, the fixed bed reactor is the flagship of the chemical industry, adding economic value to the produced materials. Figure 6 illustrates a fixed bed reactor, showing the fluid inlet and outlet, and the packed bed of particles.

It is possible to list some advantages of the fixed bed reactor in relation to similar equipment, such as simplicity and low cost of construction and maintenance. Further, in this reactor type, auxiliary equipments are not required due to the flexibility to fix the particles in the bed dispensing expenses with implantation of

Fig. 6 Scheme of a fixed bed reactor of elliptic cylindrical cross section

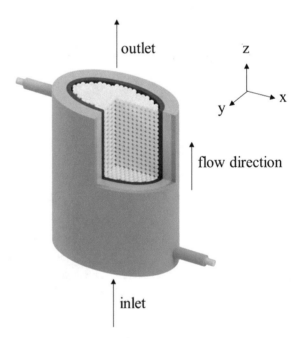

downstream separation units. It is through the use of a fixed bed reactor that ideal economic conditions can be obtained when a reaction occurs at extremely high temperatures or high pressures and employing solid catalysts [7].

Fixed-bed reactors can be scaled-up of many ways, but it is worth pointing out that the theory of similarity is of little use in this case. The usual practice is to proceed from the laboratory scale to the commercial scale gradually, passing through at least two intermediate scales: the pilot unit and the demonstration unit. Even with the use of large pilots, it is difficult to obtain exactly the same behavior of the reactor, so, a more detailed analysis is necessary. Nowadays, due to the continuous and accelerated development of information technology, and the existence and availability of industrial simulators with increasing possibility for design optimization of a specific reactor, it is not prudent to expand the scale of reactors without developing some simplest modeling [8].

The design of fixed bed reactors is quite complex because there are many variables involved in the task, such as, thermal parameters to be determined experimentally; ideal geometry to maximize process efficiency; profile of fluid velocity and temperature at the inlet region; packaging configuration of particles, which involves more than one packaging technique; angular, radial and axial measurements, which are subject to systematic and operator errors and may influence in the determination of thermal parameters; the experimental difficulty of determining the fluid-wall heat transfer coefficient, voids fraction and the fluid velocity in that region. There are yet the question related to the thermal effects of entry, since considering a flow with flat temperature profile when it would be parabolic can imply in serious errors in the determination of thermal parameters, and the heat flux from the heated section to the thermal section which can also lead to failures in the measurements of the actual temperature values and thermal parameters, especially the thermal conductivity.

3.2 Mathematical Modeling

The mathematical modeling for reactor simulation has been of fundamental importance in its design. For more sophisticated model and, greater is the number of variables influencing the process, thus, the probability of success in the simulation is higher, however, to the extent that a mathematical model aggregates more variables, the solution of model becomes difficult, even with the sophisticated CFD trading codes and very high processing power of the current computers.

The interest in dynamic simulation of chemical reactors has been boosted in recent years. This activity has helped several work purposes, namely: reactor design, reactor start and stop strategies, determination of desired operating conditions and risk for process control, control and optimization studies, etc. In the first three purposes the models to be used are more detailed, often heterogeneous, applied with the intention of obtaining more accurate results of the reactor and where the computational precision has more prominence in relation to the computational speed. For control and optimization, models are based on faster and easier solutions, generating responses that agree qualitatively with the process, which can be solved in computers with lower capacity [9].

Reliable models depend on the knowledge of how physical, chemical, and external factors affect overall system performance. The major challenge is to obtain a representative model, which maintains the essential characteristics of the system. There are several possibilities for modeling, and all of the aforementioned variables can be inserted, or a combination of most of them, but this can affect the feasibility of the solution.

The trend in fixed bed reactor modeling is the use of homogeneous or heterogeneous models depending on the nature of the phases involved in the reactor operation. The models can be considered as a pseudo-homogeneous or heterogeneous model depending on how the medium formed by the catalyst particles and reagent mixture is treated. In addition, depending on what one wants to evaluate in the reactor operation, the simulation can be one, bi or three-dimensional, either in permanent or transient regime. Next, a brief discussion about pseudo-homogeneous and heterogeneous models and their main aspects in modeling is presented.

3.2.1 Heterogeneous Models

In the two-phase model (solid + fluid), each phase has an independent dynamics in the heat transfer, requiring the definition of a parameter for each phase and another whose function is related to heat transfer between them. It is a more realistic model from the physical point of view, but its modeling has disadvantages, namely: considerably more complicated solution of the energy equations, experimental difficulty in determining the solid-fluid heat transfer coefficient required for this model and the difficulty in the punctual temperature measurement for each phase (especially in cases where $D_P \leq 3$ mm) [8].

In a heterogeneous model it is basically necessary to solve differential equations of reagent concentration, momentum and energy equations, and there are different approaches in this sense. Shafeeyan [10] qualitatively describes several possible approaches, reported by several authors in the last three decades, for the simulation of carbon dioxide adsorption in a fixed bed reactor via heterogeneous modeling. Generally, the models consider the reactant gas as an ideal gas. The flow pattern is described by the axially dispersed plug-flow model or plug-flow model. Further, it is assumed that the radial concentration gradients and, also as applied, temperature and pressure, are negligible. These assumptions have been widely accepted in many other studies [11–13]. Most of the models reported by Shafeeyan [10] include the effects of finite mass transfer rate, resulting in a theoretical representation that approaches more of a real process; use a linear approximation of the driving force to describe the gas-solid mass transfer mechanism; besides some of the models consider the effects of heat generation and heat transfer on the adsorbent bed, which may affect the adsorption indices. In the modeling of the non-isothermal adsorption processes occurring in compacted beds, it is also generally assumed that the heat transfer resistance between the gas and the solid phases is insignificant and the phases achieve thermal equilibrium instantaneously. In most models, generally the pressure drop across the bed is negligible, thus, the bed operates at constant pressure.

Amiri and Vafai [14] report a heterogeneous model in transient regime applied to incompressible fluid flow in a bed of solid spherical particles, uniform and non deformable. The heat transfer by radiation is negligible due to the high volumetric flow rate and the moderate magnitude of the operating temperatures considered in the investigation.

The governing equations are as follows:

(a) Mass conservation equation:

$$\nabla \cdot \langle V \rangle = 0 \qquad (3)$$

where v is the velocity vector.

(b) *Momentum* conservation equation:

$$\frac{\rho_f}{\varepsilon}\left[\frac{\partial \langle v \rangle}{\partial t} + \langle (v \cdot \nabla)v \rangle \right] = \frac{k_f}{K}\langle v \rangle - \frac{\rho_f F \varepsilon}{\sqrt{K}}[\langle v \rangle \cdot \langle v \rangle]J + \frac{\mu_f}{\varepsilon}\nabla^2 \langle v \rangle - \nabla \langle P \rangle^f \qquad (4)$$

(c) Energy conservation equation for the fluid phase:

$$\varepsilon \langle \rho_f \rangle^f c_f \frac{\partial \langle T_f \rangle^f}{\partial t} + \langle \rho_f \rangle^f c_f \langle v \rangle \cdot \nabla \langle T_f \rangle^f = \nabla \cdot \left[k_{feff} \cdot \nabla \langle T_f \rangle^f \right] + h_{sf}a_{sf}\left[\langle T_s \rangle^s - \langle T_f \rangle^f \right] \qquad (5)$$

(d) Energy conservation equation for the solid phase:

$$(1-\varepsilon)\rho_s c_s \frac{\partial \langle T_s \rangle^s}{\partial t} = \nabla \cdot [k_{seff} \cdot \nabla \langle T_s \rangle^s] + h_{sf}a_{sf}\left[\langle T_s \rangle^s - \langle T_f \rangle^f \right] \qquad (6)$$

In the above mentioned equations v is the velocity vector; ε is the bed porosity; F is the geometric function; μ_f is the dynamic viscosity of the fluid; $J = \frac{v_P}{|v_P|}$ is the unit vector oriented along the velocity vector; $v_P \langle P^f \rangle$ is the mean pressure given by an external meter; c_f and c_s are the specific heat of the fluid and solid phases at constant pressure, respectively; k_f and K_s are the thermal conductivity of the fluid and solid phases, respectively; ρ_f and ρ_s are the specific mass of the fluid and solid phases, respectively; and $\langle T_f \rangle^f$ and $\langle T_s \rangle^s$ are the mean intrinsic temperatures of the fluid and solid phases, respectively. For practical engineering purposes, the term $\langle (v \cdot \nabla)v \rangle$ that is responsible for the formation of the boundary layer in a short input length can be neglected.

In the right side of Eq. (4), the first term refers to the frictional resistance, the second term contemplates the inertial effects and the third term captures the viscous effects. The permeability of the bed is given by:

$$K = \frac{\varepsilon^3 D_p^2}{150(1 - \varepsilon)^2} \tag{7}$$

Based on the experimental results [15], the function F in the *momentum* conservation equation is given as follows:

$$F = \frac{1.75}{\sqrt{150\varepsilon^3}} \tag{8}$$

The porosity function generally has the following profile:

$$\varepsilon = \varepsilon_\infty \left[1 + a \exp\left(\frac{-by}{D_p} \right) \right] \tag{9}$$

In Eq. (9), a and b are constants obtained by fit to the experimental data available in the literature [16].

The specific surface area is given by [1]:

$$a_{sf} = \frac{6(1 - \varepsilon)}{D_p} \tag{10}$$

The effective thermal conductivity of the fluid is composed by two components: stagnation and dispersion, given respectively, as follows:

$$\frac{k_{feff|x}}{k_f} = \varepsilon + 0.5 \left(\frac{\rho_f u D_p}{\mu_f} \right) Pr \tag{11}$$

$$\frac{k_{feff|y}}{k_f} = \varepsilon + 0.1 \left(\frac{\rho_f u D_p}{\mu_f} \right) Pr \tag{12}$$

where u is the component of velocity vector in the x-direction e Pr is the Prandtl number.

The thermal conductivity of the solid phase has only the stagnation component, as follows:

$$k_{seff} = (1 - \varepsilon)k_s \tag{13}$$

Wakao and Kaguei [17] based on experimental results reported a correlation for the solid-fluid heat transfer coefficient, valid for $Re_p \leq 8500$, as follows:

$$h_s = k_f \left[2 + 1.1 Pr^{1.3} \frac{(\rho_f u D_p)^{0.6}}{\mu_f} \right] \Big/ D_p \tag{14}$$

Finally, Amiri and Vafai [14] report correlations for the Nusselt number written in terms of the hydraulic diameter (D_h) defined separately for each phase.

3.2.2 Pseudo-homogeneous Models

In pseudo-homogeneous models, the catalyst and reagent fluid particles are treated as a single continuous medium. It is assumed that the fluid and catalyst particle within a volume element may be characterized by a certain temperature, pressure and composition, and these quantities vary continuously with the position within the reactor [7]. The heterogeneous nature of the bed is communicated to the model only indirectly or implicitly by effective parameters.

Several pseudo-homogeneous models have been reported in the literature. Among them, we can be cited one-dimensional models, which are used preferentially for beds with low tube diameter to particle diameter ratio (D/D_p) [18]. Generally, in these models, a constant temperature is assumed along the radial position and only one transport parameter is estimated, namely, the global heat transfer coefficient.

The two-dimensional models aggregate the effects of radial and axial temperature variation. Generally, there are two parameters or more to be estimated which are the wall-bed heat transfer coefficient, and the radial and axial thermal conductivities. The diffusive term on the energy equation in the axial direction is sometimes neglected in some models, depending on the conditions of the system. This assumption presents little influence on the results and still save time and computational effort. In this case, the determination of the axial thermal conductivity is unnecessary, giving rise to the determination of a parameter so called effective thermal conductivity.

The mathematical/computational effort required to solve a specified model must be reasonable, especially when it comes to control and optimization applications [19]. In these cases, the pseudo-homogeneous models are more appropriated instead the heterogeneous models because they are simpler and predict quickly and easily the dynamic behavior of the reactor. Particularly in small particles porous beds, the difficulties involved in performing isolated measurements of fluid(s) and particle(s) temperatures make the pseudo-homogeneous approach convenient for modeling, because the phases are not distinct in this type of model.

Many researchers reported studies related to validation of pseudo-homogeneous models either by confronting experimental data, or by comparing single and two-dimensional approaches. Next, a summary of the most classic pseudo-homogeneous models will be presented.

The general energy conservation equation for a pseudo-homogeneous model is given by:

$$\frac{\partial}{\partial t}(\rho \varepsilon c_P T) + \nabla \cdot (\rho \varepsilon c_P v T) = \nabla \cdot (k \nabla T) + \dot{q} \tag{15}$$

where T is the temperature; ε is the medium porosity; ρ is the specific mass; c_p is the specific heat; k is the thermal conductivity; v is the velocity vector. In the first side of the Eq. (15), the first term is the transient term; the second is the convective term. On right side, the first term is the diffusive term and the last is the source term.

For a two-dimensional case and considering radial velocity component null, the energy conservation equation written in the cylindrical coordinates system is given as follows [8]:

$$\frac{\partial}{\partial t}\left(r \rho \varepsilon c_p T\right) + \frac{\partial}{\partial z}\left(r \rho \varepsilon c_p u_z T\right) = \frac{\partial}{\partial r}\left(rk \frac{\partial T}{\partial r}\right)$$
$$+ \frac{\partial}{\partial \theta}\left(\frac{1}{r} k \frac{\partial T}{\partial \theta}\right) + \frac{\partial}{\partial z}\left(rk \frac{\partial T}{\partial z}\right) + \dot{q} r \tag{16}$$

where r and z are radial and axial coordinates, respectively; θ is the angular coordinate and u_z is the fluid velocity at the reactor inlet.

It is practically impossible to obtain an analytical solution of the Eq. (15) because of the non-linearity of its terms. Thus, some physical considerations are usually made in order to eliminate some terms that may have less relevance in the phenomenon.

The basic formulation of the energy equation applied to gas-liquid flow in a fixed bed of circular cross section filled with particle, after some physical considerations will be [20]:

$$\left(Gc_{P_g} + Lc_{P_l}\right)\frac{\partial T}{\partial z} = k_r \left[\frac{1}{r}\frac{\partial}{\partial r} r \frac{\partial T}{\partial r}\right] + k_z \frac{\partial^2 T}{\partial z^2} \tag{17}$$

where G is the gas mass flow, L is the liquid mass flow, c_{pg} is the gas specific heat; c_{pl} is the liquid specific heat; k_r is the radial thermal conductivity, and k_z is the axial thermal conductivity.

Equation (17) can have different solutions according to the physical considerations made [8, 21–23]. The axial thermal conductivity; and (k_z) is sometimes neglected, thus, reducing the number of parameters in the solution [20].

Analytical solutions are usually processed by some integral transform technique or by separating variables. If the task of simulating fixed-bed reactor wants to incorporate more embracing and realistic occurrences, one must consider the use of numerical methodologies of solution.

The literature reports variants of the pseudo-homogeneous model, which originally contemplates three parameters, namely, k_z, k_r and h_w, which are the axial thermal conductivity, radial thermal conductivity and the heat transfer coefficient in the bed-wall, respectively (Table 1).

One can still cite the model reported by Oliveira [8], which contain two parameters (k_r and k_z) and considers gas flow only. In this case, the energy equation takes the form:

$$Gc_P \frac{\partial T}{\partial z} = \frac{k_r}{r} \frac{\partial}{\partial r}\left(r \frac{\partial T}{\partial r}\right) + k_z \frac{\partial}{\partial z}\left(\frac{\partial T}{\partial z}\right) \tag{23}$$

The boundary conditions used by Oliveira [8] were the same reported by Dixon et al. [22] (model 5), except for the boundary condition at the reactor inlet which, in turn, resembles that of model 4 [23], but with a more complex form, as follows:

$$T_0^*(r) = \frac{(T(r) - T_w)}{(T_0 - T_w)} = \sum_{n=1}^{\infty} A_n J_0\left(a_n \frac{r}{R}\right) \tag{24}$$

where:

$$A_n = \frac{2}{R^2} \frac{\int_0^R r J_0\left(a_n \frac{r}{R}\right) T_0^*(r) dr}{\left[\left(\frac{a_n}{Bi}\right)^2 + 1\right] J_1^2(a_n)} \tag{25}$$

In Eq. (25) J_0 and J_1 are the Bessel functions of first kind of order 0 and 1, respectively.

The governing energy equation was solved analytically by the technique of separation of variables through the use of Bessel functions and their orthogonality properties. The exact solution of Eq. (23) is given as follows:

$$T^*(r,z) = \sum_{n=1}^{\infty} A_n J_0(\lambda_n r) \exp \frac{Gc_P z}{2k_z}\left(1 - \sqrt{1 + 4\left(\frac{\lambda_n}{Gc_P}\right)^2 k_r k_z}\right) \tag{26}$$

From the pseudo-homogeneous models exposed, it can be seen that on the surface ($r = R$) there can be a boundary condition with constant temperature or convection, the term of axial dispersion [$k_z \frac{\partial^2 T}{\partial z^2}$] may or may not be present depending on the simplification of the model, in order to avoid the experimental determination of the parameter k_z. It is also possible to use a constant temperature or a parabolic temperature profile at the reactor inlet ($z = 0$). Model 5 better capture the occurrences in the system, since that T_0 is obtained more carefully considering

Table 1 Different pseudo-homogeneous model for heat transfer in fixed bed reactor

Model	Basic equation in pseudo-homogeneous formulation	Boundary and symmetry conditions	Source
1	$\left(Gc_{p_g} + Lc_{p_h}\right)\dfrac{\partial T}{\partial z} = k_r\left[\dfrac{1}{r}\dfrac{\partial}{\partial r}r\dfrac{\partial T}{\partial r}\right]$ (18)	$r = R: T = T_w$ $r = 0: \dfrac{\partial T}{\partial r} = 0$ $z = 0: T = T_0$	[20]
2	$\left(Gc_{p_g} + Lc_{p_h}\right)\dfrac{\partial T}{\partial z} = k_r\left[\dfrac{1}{r}\dfrac{\partial}{\partial r}r\dfrac{\partial T}{\partial r}\right] + k_a\dfrac{\partial^2 T}{\partial z^2}$ (19)	$r = R: T = T_w$ $r = 0: \dfrac{\partial T}{\partial r} = 0$ $z \to +\alpha: T = T_w$ $z = 0: T = T_0$	[20]
3	$\left(Gc_{p_g} + Lc_{p_h}\right)\dfrac{\partial T}{\partial z} = k_r\left[\dfrac{1}{r}\dfrac{\partial}{\partial r}r\dfrac{\partial T}{\partial r}\right]$ (20)	$r = R: -k_r\dfrac{\partial T}{\partial r} = h_w(T_{r=R} - T_w)$ $r = 0: \dfrac{\partial T}{\partial r} = 0$ $z = 0: T = T_0$	[20]
4	$\left(Gc_{p_g} + Lc_{p_h}\right)\dfrac{\partial T}{\partial z} = k_r\left[\dfrac{1}{r}\dfrac{\partial}{\partial r}r\dfrac{\partial T}{\partial r}\right]$ (21)	$r = R: -k_r\dfrac{\partial T}{\partial r} = h_w(T_{r=R} - T_w)$ $r = 0: \dfrac{\partial T}{\partial r} = 0$ $z = 0: \dfrac{T - T_w}{T - T_0} = 1 - A\left(\dfrac{r}{R^2}\right)$	[23]
5	$\left(Gc_{p_g} + Lc_{p_h}\right)\dfrac{\partial T}{\partial z} = \left[\dfrac{1}{r}\dfrac{\partial}{\partial r}r\dfrac{\partial T}{\partial r}\right] + k_a\dfrac{\partial^2 T}{\partial z^2}$ (22)	$r = R: -k_r\dfrac{\partial T}{\partial r} = h_w(T_{r=R} - T_w)$ $r = 0: \dfrac{\partial T}{\partial r} = 0$ $z \to +\infty: T = T_w$ T_0 is determined by energy balances in the transition to the thermal section	[22]

occurrences upstream of the thermal inlet through an energy balance in the undisturbed section.

Giudici [24] analyzed several of these models with parameter $k_r - h_w - k_z$ or $k_r - h_w$, with or without convective boundary condition at the reactor surface. This author verified that the parameter k_z is very sensitive, so that small variations in the Reynolds number promote very different results for this parameter, being very difficult to explain it. Then, the author discourages the use of this parameter in the modeling.

In the models reported in Table 1, statistical techniques such as least squares errors, for example, are of great value, in order to assist in the determination of the thermal parameters, since they help to infer about the quality of the parameter estimation in nonlinear models and their effects on the performance of the model.

Silva et al. [25] report a transient and three-dimensional model considering thermophysical properties as function of the temperature or position in a fixed-bed reactor of elliptic-cylindrical geometry and a curved temperature profile of the gas phase at the reactor inlet. This model will be detailed in the next section.

4 Transient Heat Transfer in Fixed-Bed Elliptic-Cylindrical Reactor

4.1 Physical Problem and Geometry

The system under study is shown in Fig. 7 (elliptic-cylindrical reactor), where the sketch of the fixed bed filled with particles is observed. The reactor bed is percolated by a heated fluid (fluid 1). Thus, fluid exchanges heat with the particles and the assembly is cooled at the surface by a cooling fluid (fluid 2) at a temperature lower than the inlet fluid temperature. The flow of the fluid 1 is in a laminar regime.

Sometimes a given physical problem has a certain geometry which is complicated to implement the boundary conditions in Cartesian coordinates. Then, the representation of the boundary conditions in most physical problems requires that the values of a function (or its derivatives) be specified in the form of curves or surfaces (spheres, cylinders, etc.). Therefore, the geometric form of a physical system under study has pointed to the use of a given coordinate system that best fits this geometry and, consequently, will lead to a greater efficiency in the validation of the predicted results [26].

When it is desired to work with a curvilinear coordinate system, such as the elliptic-cylindrical, changes in the variables are required. These changes result from the geometric form of the system under consideration. The relationships between the Cartesian coordinate systems (x, y, z) and the elliptic-cylindrical coordinate system (τ, φ, z) are given as follow [26]:

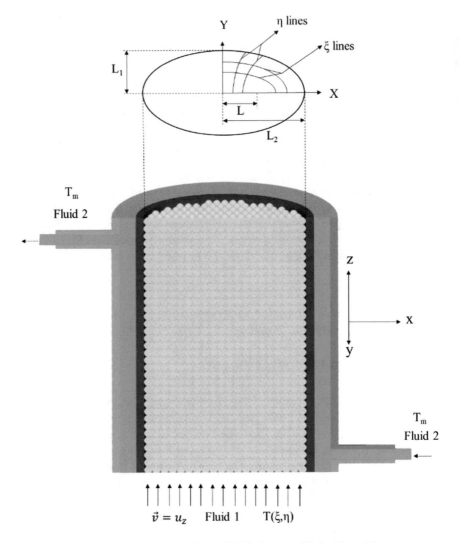

Fig. 7 Representation of a fixed bed elliptic cylindrical reactor filled with particles

$$x = L \cos h\tau \, \text{sen} \varphi \qquad (27)$$

$$x = L \cos h\tau \, \text{sen} \varphi \qquad (28)$$

$$z = z \qquad (29)$$

In Fig. 8, the elliptic cross-section of the reactor can be observed. The parameter L_1 is the smaller half-axis, L_2 is the largest half-axis and L is the focal length of the ellipse. Thus, considering the following variables:

Fig. 8 Representation of the
variables applied in the
Eq. (65)

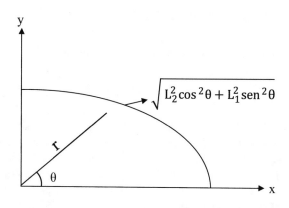

$$\xi = \cosh \tau \tag{30}$$

$$\eta = \cos \varphi \tag{31}$$

$$z = z \tag{32}$$

and with direct substitution of these variables in Eqs. (27)–(29), the relations
between the two coordinate systems are given as follows [27, 28]:

$$x = L\xi\eta \tag{33}$$

$$y = L \sqrt{(1 - \eta^2)(\xi^2 - 1)} \tag{34}$$

$$z = z \tag{35}$$

where $1 \leq \xi \leq L_2/L$, $0 \leq \eta \leq 1$ and $0 \leq z \leq H$. The line $\xi = 1$ is equivalent to
the straightline segment with $(0, 0)$ and $(L, 0)$ ends. The line $\xi = L_2/L$ represents
the curve line of the ellipse; the line $\eta = 0$ is equivalent to the line segment with the
ends in the points $(0, 0)$ and $(0, L_1)$; and the line s is equivalent to the line segment
with the ends in the points $(L, 0)$ and $(L_2, 0)$. The lines ξ with constant value are
ellipses of the same center, except for $\xi = 1$. Further, the lines η with constant
value, except for $\eta = 0$ and $\eta = 1$, are half-branches of hyperbole. Herein, this
physical problem will be treated only in the first quadrant of the ellipse, due to the
symmetry with respect to the x and y axes.

4.2 Governing Equations

The general energy conservation equation is given by:

$$\frac{\partial}{\partial t}(\rho \varepsilon \, c_P T) + \nabla \cdot (\rho \varepsilon c_P v T) = \frac{DP}{Dt} + \nabla \cdot (k \nabla T) + \mu \Psi + \dot{q} \tag{36}$$

In Eq. (36) ρ is the specific mass, c_p is the specific heat and k is the thermal conductivity of the pseudo-phase; v is the velocity vector; ε is the porosity of the bed; is the substantive derivative of pressure; $\mu \Psi$ is the viscous dissipation term and $\dot{q}$ is the term of internal energy generation in the system.

The general energy conservation equation (Eq. 36) was applied to a pseudo-homogeneous system, where the particulate and fluid phases are considered with a single-phase. In this approach, both the phases are at same temperature at any point of contact (local thermal equilibrium).

Applying the methodology described by [27–29], Eq. (36) can be written for any coordinate system, as follows:

$$\frac{\partial}{\partial t}\left(\rho \varepsilon \, c_p \frac{T}{J}\right) + \frac{\partial}{\partial \xi}\left(\rho \varepsilon \, c_p \, \tilde{U} T\right) + \frac{\partial}{\partial \eta}\left(\rho \varepsilon \, c_p \tilde{V} T\right) + \frac{\partial}{\partial z}\left(\rho \varepsilon \, c_p \widetilde{W} T\right)$$

$$= \frac{\partial}{\partial \xi}\left(\alpha_{11} J \, k \frac{\partial T}{\partial \xi} + \alpha_{12} J \, k \frac{\partial T}{\partial \eta} + \alpha_{13} J \, k \frac{\partial T}{\partial z}\right)$$

$$+ \frac{\partial}{\partial \eta}\left(\alpha_{21} J \, k \frac{\partial T}{\partial \xi} + \alpha_{22} J \, k \frac{\partial T}{\partial \eta} + \alpha_{23} J \, k \frac{\partial T}{\partial z}\right)$$

$$+ \frac{\partial}{\partial z}\left(\alpha_{31} J \, k \frac{\partial T}{\partial \xi} + \alpha_{32} J \, k \frac{\partial T}{\partial \eta} + \alpha_{33} J \, k \frac{\partial T}{\partial z}\right)$$

$$+ \frac{DP}{Dt} + \frac{\mu \Psi}{J} + \frac{\dot{q}^\Phi}{J} \tag{37}$$

where J is the Jacobian of the transformation of the Cartesian system to the transformed plane.

The temperature gradient, the areas of heat exchange and the volume of the infinitesimal element in the new coordinate system can be obtained using the mathematical relations provided by [27, 29–31], as follows:

- Differential volume:

$$dV = \frac{1}{\widehat{U}\,\widehat{V}\,\widehat{W}} d\xi d\eta dz \tag{38}$$

- Differential areas for heat flux:

$$dS_\xi = \frac{1}{\widehat{V}\,\widehat{W}} d\eta dz \tag{39}$$

$$dS_\eta = \frac{1}{\widehat{U}\widehat{W}}d\xi dz \tag{40}$$

$$dS_z = \frac{1}{\widehat{U}\widehat{V}}d\xi d\eta \tag{41}$$

- Temperature gradient

$$\nabla T = \left(\widehat{U}\frac{\partial T}{\partial \xi} ; \widehat{V}\frac{\partial T}{\partial \eta} ; \widehat{W}\frac{\partial T}{\partial z} \right) \tag{42}$$

In the Eqs. (38)–(42), the metric coefficients $\widehat{U}$, $\widehat{V}$ and $\widehat{W}$, whose product generates as a result the Jacobian of the transformation (J), are given by:

$$\frac{1}{\widehat{U}^2} = \left(\frac{\partial x}{\partial \xi}\right)^2 + \left(\frac{\partial y}{\partial \xi}\right)^2 + \left(\frac{\partial z}{\partial \xi}\right)^2 \tag{43}$$

$$\frac{1}{\widehat{V}^2} = \left(\frac{\partial x}{\partial \eta}\right)^2 + \left(\frac{\partial y}{\partial \eta}\right)^2 + \left(\frac{\partial z}{\partial \eta}\right)^2 \tag{44}$$

$$\frac{1}{\widehat{W}^2} = \left(\frac{\partial x}{\partial z}\right)^2 + \left(\frac{\partial y}{\partial z}\right)^2 + \left(\frac{\partial z}{\partial z}\right)^2 \tag{45}$$

The coefficients α_{ij}, are given by the following mathematical relations:

$$\alpha_{11} = \frac{a'}{J^2} ; \tag{46}$$

$$\alpha_{22} = \frac{b'}{J^2} ; \tag{47}$$

$$\alpha_{12} = \alpha_{21} = \frac{d'}{J^2} ; \tag{48}$$

$$\alpha_{13} = \alpha_{31} = \frac{e'}{J^2} ; \tag{49}$$

$$\alpha_{23} = \alpha_{32} = \frac{f'}{J^2} \tag{50}$$

with:

$$a' = \left(\frac{\partial \xi}{\partial x}\right)^2 + \left(\frac{\partial \xi}{\partial y}\right)^2 + \left(\frac{\partial \xi}{\partial z}\right)^2 \tag{51}$$

$$b' = \left(\frac{\partial \eta}{\partial x}\right)^2 + \left(\frac{\partial \eta}{\partial y}\right)^2 + \left(\frac{\partial \eta}{\partial z}\right)^2 \tag{52}$$

$$c' = \left(\frac{\partial z}{\partial x}\right)^2 + \left(\frac{\partial z}{\partial y}\right)^2 + \left(\frac{\partial z}{\partial z}\right)^2 \tag{53}$$

$$d' = \left(\frac{\partial \xi}{\partial x}\frac{\partial \eta}{\partial x}\right) + \left(\frac{\partial \xi}{\partial y}\frac{\partial \eta}{\partial y}\right) + \left(\frac{\partial \xi}{\partial z}\frac{\partial \eta}{\partial z}\right) \tag{54}$$

$$e' = \left(\frac{\partial z}{\partial x}\frac{\partial \xi}{\partial x}\right) + \left(\frac{\partial z}{\partial y}\frac{\partial \xi}{\partial y}\right) + \left(\frac{\partial z}{\partial z}\frac{\partial \xi}{\partial z}\right) \tag{55}$$

$$f' = \left(\frac{\partial z}{\partial x}\frac{\partial \eta}{\partial x}\right) + \left(\frac{\partial z}{\partial y}\frac{\partial \eta}{\partial y}\right) + \left(\frac{\partial z}{\partial z}\frac{\partial \eta}{\partial z}\right) \tag{56}$$

From the Eqs. (51)–(56) it can be verified that, due to the orthogonality of the system, the coefficients d', e' and f' are null, so consequently the coefficients α_{ij} with $i \neq j$ are equal to zero.

In the Eq. (37), the variables $\widetilde{U}$, $\widetilde{V}$ and $\widetilde{W}$ are the contravariant components of the velocity vector, calculated as follows:

$$\widetilde{U} = \frac{1}{J}\left(\frac{\partial \xi}{\partial t} + u\frac{\partial \xi}{\partial x} + v\frac{\partial \xi}{\partial y} + w\frac{\partial \xi}{\partial z}\right) \tag{57}$$

$$\widetilde{V} = \frac{1}{J}\left(\frac{\partial \eta}{\partial t} + u\frac{\partial \eta}{\partial x} + v\frac{\partial \eta}{\partial y} + w\frac{\partial \eta}{\partial z}\right) \tag{58}$$

$$\widetilde{W} = \frac{1}{J}\left(\frac{\partial z}{\partial t} + u\frac{\partial z}{\partial x} + v\frac{\partial z}{\partial y} + w\frac{\partial z}{\partial z}\right) \tag{59}$$

The mathematical relations for converting the velocity vector components in the Cartesian coordinates system for the curvilinear coordinates system are given by:

$$u_\xi = \widehat{U}\left(u\frac{\partial x}{\partial \xi} + v\frac{\partial y}{\partial \xi} + w\frac{\partial z}{\partial \xi}\right) \tag{60}$$

$$u_\eta = \widehat{V}\left(u\frac{\partial x}{\partial \eta} + v\frac{\partial y}{\partial \eta} + w\frac{\partial z}{\partial \eta}\right) \tag{61}$$

$$u_z = \widehat{W}\left(u\frac{\partial x}{\partial z} + v\frac{\partial y}{\partial z} + w\frac{\partial z}{\partial z}\right) \tag{62}$$

By using Eqs. (36)–(62) and after of an adequate mathematical treatment in Eq. (37), we obtain the general energy conservation equation in the elliptic-cylindrical coordinate system. However, for an appropriate mathematical modeling, the following physical considerations have been established:

  (i) The system operates in the transient regime;
 (ii) The thermophysical properties of the phases change with the bed temperature;
(iii) The porosity varies with the radial position of the bed;
 (iv) There is an axial symmetry of temperature with respect to the x and y axes;
  (v) There is no chemical reaction, therefore, the internal energy generation term is null;
 (vi) The superficial velocity of the fluid is constant at any position within the reactor along the z-coordinate and null in the directions ξ and η;
(vii) The superficial velocity of the fluid is less than the sound speed;
(viii) The wall thickness of the reactor is negligible.
 (ix) Viscous effects are insignificant;
  (x) A convective boundary condition is considered at the surface of the reactor.

Then, after the mathematical manipulations and the application of the physical considerations, the energy conservation equation in the elliptic-cylindrical coordinate system assumes the following form:

$$\frac{\partial}{\partial t}\left[\frac{\rho\varepsilon c_p L^2(\xi^2 - \eta^2)T}{\sqrt{(\xi^2 - 1)(1 - \eta^2)}}\right] + \frac{\partial}{\partial z}\left[\frac{\rho\varepsilon c_p L^2(\xi^2 - \eta^2)u_z T}{\sqrt{(\xi^2 - 1)(1 - \eta^2)}}\right]$$

$$= \frac{\partial}{\partial \xi}\left[\sqrt{\frac{\xi^2 - 1}{1 - \eta^2}}k\frac{\partial T}{\partial \xi}\right] + \frac{\partial}{\partial \eta}\left[\sqrt{\frac{1 - \eta^2}{\xi^2 - 1}}k\frac{\partial T}{\partial \eta}\right] + \frac{\partial}{\partial z}\left[\frac{L^2(\xi^2 - \eta^2)k}{\sqrt{(\xi^2 - 1)(1 - \eta^2)}}\frac{\partial T}{\partial z}\right]$$

$$\tag{63}$$

The first term of the Eq. (63) is the transient accumulation term and the second is the convective term in the axial direction. After equality, there are three terms representing the heat diffusive transport in the directions ξ, η and z, respectively.

4.3   *Initial, Symmetry and Boundary Conditions*

(a) Initial condition:

$$T(\xi, \eta, z, t = 0) = T_S, \ z \neq 0 \tag{64}$$

(b) Boundary condition at the reactor inlet:

$$T(\xi, \eta, z = 0, t) = T_M - \frac{2 \cdot u_z R^2}{\alpha} \left(\frac{dT_M}{dz}\right) \left[\frac{3}{16} + \frac{1}{16}\left(\frac{r}{R}\right)^4 - \frac{1}{4}\left(\frac{r}{R}\right)^2\right] \tag{65}$$

(c) Boundary condition at the reactor wall:

$$\frac{-k_f}{L} \sqrt{\frac{\xi_n^2 - 1}{\xi_n^2 - \eta_P^2}} \left.\frac{\partial T}{\partial \xi}\right|_{\xi = \frac{L_2}{L}} = h_w \left[T\left(\xi = \frac{L_2}{L}, \eta, z, t\right) - T_m\right] \tag{66}$$

(d) Boundary condition at the reactor outlet:

$$\frac{\partial T}{\partial z}(\xi, \eta, z = H, t) = 0 \tag{67}$$

(e) Condition in the symmetry planes:

$$\frac{\partial T}{\partial z}(\xi = 1, \eta, z, t) = 0 \tag{68}$$

$$\frac{\partial T}{\partial z}(\xi, \eta = 0, z, t) = 0 \tag{69}$$

$$\frac{\partial T}{\partial z}(\xi, \eta = 1, z, t) = 0 \tag{70}$$

The Eq. (65) is equivalent to a parabolic temperature profile in a circular cylinder [32] adapted for an elliptical cross-section, considering R as half of the hydraulic diameter. It is noteworthy that this methodology is flexible in such a way that it is valid to simulate cases from a circular cylinder, passing through a rectangular channel to an elliptical cylinder. In the Eq. (65), α is the thermal diffusivity and u_z is the axial velocity of the fluid, and dT_M/dz is the mean axial temperature gradient of the gas phase at z = 0 m, given by [8]:

$$\frac{dT_M}{dz} = \frac{-(T_M - T_m)}{\frac{11}{48} \cdot \frac{u_z R^2}{\alpha}} \tag{71}$$

in which T_M is the average temperature at the inlet section of the reactor and T_m is the temperature on the reactor surface at the inlet. Further, r is the radial coordinate varying as follows: $0 \leq r \leq \sqrt{L_2 \cos^2 \theta + L_1 \text{sen}^2 \theta}$ where θ is the angle ranging in the interval $0 \leq \theta \leq 90°$ counterclockwise [10] (see Fig. 8).

4.4   Numerical Treatment

For the solution of the energy conservation equation in the elliptic-cylindrical coordinate system, the numerical method of finite volumes was applied, which consists of dividing the domain under study into several subdomains called control volumes. It is assumed that if the conservation equation is valid for the entire domain, it will also be valid for any subdomain. The control volumes are referenced by nodal points located in their centroids, in which the temperature is determined. In this context, the conservation equation is integrated for each control volume in space and time, producing as a result a linear algebraic equation for each control volume. The set of all equations generated with the proper application of the boundary conditions form a system of possible linear algebraic equations determined to be solved with some iterative or direct method.

Figure 9 shows a representative control volume element appropriated to the problem domain. The central nodal point P as well as the central nodal points of the neighboring volumes N, S, E, W, F, and T are illustrated also. The limits between the faces of the control volumes are referenced by lowercase letters, namely n, s, e, w, f and t.

By integrating the Eq. (63) in volume and time, and using the WUDS interpolation scheme for the convective terms and a fully implicit formulation for the evaluation of the solution over time, the following linear algebraic equation was obtained:

$$A_P T_P = A_E T_E + A_W T_W + A_N T_N + A_S T_S + A_F T_F + A_T T_T + A_P^0 T_P^0 + B \tag{72}$$

where the coefficients A_k of this equation are given by:

$$A_E = \sqrt{\frac{1 - \eta_e^2}{\xi_e^2 - 1}} \frac{k_e \beta_e \Delta\xi \Delta z}{\delta\eta_e} \tag{73}$$

$$A_W = \sqrt{\frac{1 - \eta_w^2}{\xi_w^2 - 1}} \frac{k_w \beta_w \Delta\xi \Delta z}{\delta\eta_w} \tag{74}$$

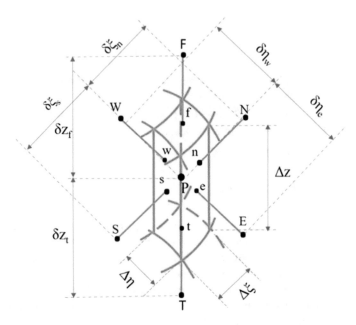

Fig. 9 Representation of a three-dimensional elemental control volume in the elliptic-cylindrical coordinate system

$$A_S = \sqrt{\frac{\xi_s^2 - 1}{1 - \eta_s^2}} \frac{k_s \beta_s \Delta\eta\Delta z}{\delta\eta_s} \tag{75}$$

$$A_N = \left\{ \begin{array}{cc} \sqrt{\frac{\xi_n^2}{1-\eta_n^2}} \frac{K_n \beta_n \Delta\eta\Delta z}{\delta\eta_n} & \text{for internal points} \\ 0, & \text{for the border points} \end{array} \right. \tag{76}$$

$$A_F = \frac{L^2(\xi_f^2 - \eta_f^2)}{\sqrt{\left(\xi_f^2 - 1\right)\left(1 - \eta_f^2\right)}} \left[(0.5 + \alpha_f)\rho_f \varepsilon_f c_{Pf} u_z + \frac{\beta_f k_f}{\delta z_f} \right] \Delta\xi\Delta\eta \tag{77}$$

$$A_T = \frac{L^2(\xi_t^2 - \eta_t^2)}{\sqrt{(\xi_t^2 - 1)\left(1 - \eta_t^2\right)}} \left[(0.5 - \alpha_t)\rho_t \varepsilon_t c_{Pt} u_z + \frac{\beta_t k_t}{\delta z_t} \right] \Delta\xi\Delta\eta \tag{78}$$

$$A_P^0 = \frac{L^2(\xi^2 - \eta^2)}{\sqrt{(\xi^2 - 1)(1 - \eta^2)}} \frac{\rho^0 c_P^0 \varepsilon^0 \Delta\xi\Delta\eta\Delta z}{\Delta t} \tag{79}$$

$$A_P = A_E + A_W + A_N + A_S + A_F^* + A_T^* +$$

$$+ \frac{L^2\left(\xi_P^2 - \eta_P^2\right)}{\sqrt{\left(\xi_P^2 - 1\right)\left(1 - \eta_P^2\right)}} \frac{\rho \, c_P \varepsilon \Delta\xi \Delta\eta \Delta z}{\Delta t}$$

$$+ A_P^0 + \overline{SM} \tag{80}$$

with

$$A_F^* = \frac{L^2\left(\xi_f^2 - \eta_f^2\right)}{\sqrt{\left(\xi_f^2 - 1\right)\left(1 - \eta_f^2\right)}} \left[(\alpha_f + 0.5)\rho_f \varepsilon_f c_{Pf} u_z + \frac{\beta_f k_f}{\delta z_f}\right] \Delta\xi \Delta\eta \tag{81}$$

$$A_T^* = \frac{L^2\left(\xi_t^2 - \eta_t^2\right)}{\sqrt{\left(\xi_t^2 - 1\right)\left(1 - \eta_t^2\right)}} \left[(\alpha_t - 0.5)\rho_t \varepsilon_t c_{Pt} u_z + \frac{\beta_t k_t}{\delta z_t}\right] \Delta\xi \Delta\eta \tag{82}$$

$$\overline{SM} = \begin{cases} \dfrac{L\Delta\eta\Delta z \sqrt{\frac{(\xi_n - \eta_P)}{(1 - \eta_P)}}}{\dfrac{\left(1 + \frac{kU}{h_w \Delta\xi_n}\right)}{\left(\frac{kU}{\Delta\xi_n}\right)}}, & \text{for internal points} \\[20pt] 0, & \text{for internal points} \end{cases} \tag{83}$$

$$B = \begin{cases} \dfrac{T_m L\Delta\eta\Delta z \sqrt{\frac{(\xi_n - \eta_P)}{(1 - \eta_P)}}}{\left[\dfrac{\left(1 + \frac{kU}{h_w \delta\xi_n}\right)}{\left(\frac{kU}{\delta\xi_n}\right)}\right]}, & \text{for the border points} \\[20pt] 0, & \text{for internal points} \end{cases} \tag{84}$$

Figure 10 illustrates the nodal point close to the surface of the reactor.

The coefficients α and β that appear in terms of the coefficients A_k are weighting constants used in the interpolation functions necessary to evaluate the convective and diffusive terms at the faces of each control volume [28, 33]. These parameters are calculated by:

$$\alpha = \frac{Pe^2}{10 + 2Pe^2} \tag{85}$$

$$\beta = \frac{1 + 0.005Pe^2}{1 + 0.05Pe^2} \tag{86}$$

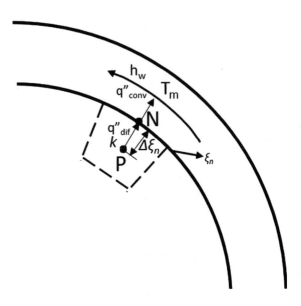

where Pe is the Peclet number, which physically represents the ratio of the convective to the diffusive flux in the coordinate direction considered. It will only make sense to determine the Peclet number in the z-direction. Thus, in this direction, it is calculated as follows:

$$Pe = \frac{\rho \varepsilon u_z c_p \delta z}{k} \tag{87}$$

The following function was used to calculate the properties at the interfaces of control volumes:

$$k_i^\Phi = \left(\frac{1 - \hat{f}_i}{k_P^\Phi} + \frac{\hat{f}_i}{k_j^\Phi} \right)^{-1} \tag{88}$$

This condition can be visualized in Fig. 11, where P is the index for the nodal point in question, $i = n, s, f, t,$ or w and $j = N, S, F, T, E$ or W. When i assumes a minuscule index, j—representing a j-th neighboring nodal point associated with that face—assumes its respective capital index. Thus, we define $\hat{f}_i$ as an interpolation factor as follows:

$$\hat{f}_i = \frac{(\delta d)_i^+}{(\delta d)_i} \tag{89}$$

where δd is given according Fig. 11.

Fig. 11 Schematic showing
the position of the i-th
interface between the points P
and a j-th neighbor nodal
point

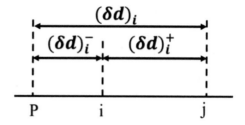

Fig. 12 Illustration of the
energy balance for
temperature estimation at
symmetry points in $\eta = 0$

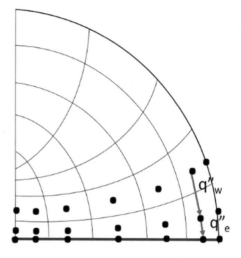

The symmetry points are not present in the set of discretized equations. Thus, after that the system of equations is solved, the temperature value in the symmetry points is calculated by a thermal balance in these points. From this analysis, we obtain that the diffusive heat flux that reaches a specific point of symmetry originated from its internal neighbor point is equal to that which arrives at this neighbor point originated from the next neighbor point, in the same coordinate direction (Figs. 12, 13 and 14).

Thus, in the region of symmetry for $\eta = 0$, we have that (Fig. 12):

$$q''_w = q''_e \Rightarrow \left(-\frac{k}{L}\sqrt{\frac{1-\eta^2}{\xi^2-\eta^2}}\frac{\partial T}{\partial \eta} \right)\Bigg|_e = \left(-\frac{k}{L}\sqrt{\frac{1-\eta^2}{\xi^2-\eta^2}}\frac{\partial T}{\partial \eta} \right)\Bigg|_w \qquad (90)$$

After discretization of the Eq. (90) the temperature at any symmetry point in this region is given as follows:

Fig. 13 Illustration of the
energy balance for
temperature estimation at
symmetry points in $\eta = 1$

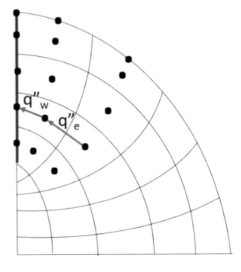

Fig. 14 Illustration of the
energy balance for
temperature estimation at
symmetry points in $\xi = 1$

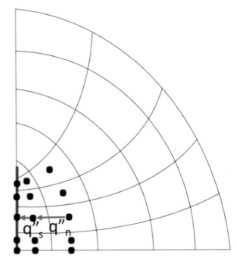

$$T_E = \left(1 + \frac{\frac{k_w}{\delta\eta_w}\sqrt{\frac{1-\eta_w^2}{\xi^2-\eta_w^2}}}{\frac{k_e}{\delta\eta_e}\sqrt{\frac{1-\eta_e^2}{\xi^2-\eta_e^2}}}\right)T_P - \left(\frac{\frac{k_w}{\delta\eta_w}\sqrt{\frac{1-\eta_w^2}{\xi^2-\eta_w^2}}}{\frac{k_e}{\delta\eta_e}\sqrt{\frac{1-\eta_e^2}{\xi^2-\eta_e^2}}}\right)T_W \qquad (91)$$

Similarly, for the points at $\eta = 1$, we have as result the following energy balance
(Fig. 13):

$$q''_e = q''_w \Rightarrow \left(-\frac{k}{L} \sqrt{\frac{1-\eta^2}{\xi^2-\eta^2}} \frac{\partial T}{\partial \eta} \right)\Bigg|_w = \left(-\frac{k}{L} \sqrt{\frac{1-\eta^2}{\xi^2-\eta^2}} \frac{\partial T}{\partial \eta} \right)\Bigg|_e \qquad (92)$$

After discretization, the temperature in these points can be determined as follows:

$$T_W = \left(1 + \frac{\frac{k_e}{\delta\eta_e} \sqrt{\frac{1-\eta_e^2}{\xi^2-\eta_e^2}}}{\frac{k_w}{\delta\eta_w} \sqrt{\frac{1-\eta_w^2}{\xi^2-\eta_w^2}}} \right) T_P - \left(\frac{\frac{k_e}{\delta\eta_e} \sqrt{\frac{1-\eta_e^2}{\xi^2-\eta_e^2}}}{\frac{k_w}{\delta\eta_w} \sqrt{\frac{1-\eta_w^2}{\xi^2-\eta_w^2}}} \right) T_E \qquad (93)$$

For the points of symmetry in which $\eta = 1$, we can write (Fig. 14):

$$q''_n = q''_s \Rightarrow \left(-\frac{k}{L} \sqrt{\frac{\xi^2-1}{\xi^2-\eta^2}} \frac{\partial T}{\partial \xi} \right)\Bigg|_n = \left(-\frac{k}{L} \sqrt{\frac{\xi^2-1}{\xi^2-\eta^2}} \frac{\partial T}{\partial \xi} \right)\Bigg|_s \qquad (94)$$

In the discretized form, we obtain the temperature in the symmetry points as follows:

$$T_S = \left(1 + \frac{\frac{k_n}{\delta\eta_n} \sqrt{\frac{\xi_n^2-1}{\xi_n^2-\eta^2}}}{\frac{k_s}{\delta\eta_s} \sqrt{\frac{\xi_s^2-1}{\xi_s^2-\eta^2}}} \right) T_P - \left(\frac{\frac{k_n}{\delta\eta_n} \sqrt{\frac{\xi_n^2-1}{\xi_n^2-\eta^2}}}{\frac{k_s}{\delta\eta_s} \sqrt{\frac{\xi_s^2-1}{\xi_s^2-\eta^2}}} \right) T_N \qquad (95)$$

Regarding the surface points, all temperature involved are determined based on the thermal balance in this region, whose representation is shown in Fig. 10.

In practice it is considered that the diffusive heat flux from the fluid to the bed wall (internal side) is equal to the convective heat flux at the same wall (external side) to the refrigerant fluid (Eq. 66). Thus, the temperature in the nth-face is given by:

$$T_N = \frac{T_m + T_P\left(\frac{K\hat{u}}{h_w \delta\xi_n}\right)}{1 + \frac{K\hat{u}}{h_w \delta\xi_n}} \qquad (96)$$

5 Application

5.1 Simulation Data

Here, it is considered the flow of heated air by percolating a bed of spherical particles in local thermal equilibrium condition. The reactor wall is cooled by water to the temperature T_m.

The geometric parameters used in the simulation were $L_2 = 0.1$ m, $L_1 = 0.05$ m and $H = 0.194$ m. The values of thermophysical parameters were $h_w = 2$ W/m^{2}K, $u_z = 0.11417$ m/s, $T_M = 70$ °C and $T_m = 30$ °C. The thermal conductivity k, the specific mass, the viscosity, and specific heat of the air, are as follows [34, 35]:

$$k = 2.425 \times 10^{-2} + 7.889 \times 10^{-5}T - 1.907 \times 10^{-8}T^2 - 8.570 \times 10^{-12}T^3 \left[\frac{W}{m\ K} \right]$$
(97)

$$\rho = \frac{P_{atm}M_{ar}}{R_{ar}T} \left[\frac{kg}{m^3} \right]$$
(98)

$$\mu = 1.691 \times 10^{-5} + 4.984. \times 10^{-8}T - 3.187 \times 10^{-11}T^2 - 1.3196 \times 10^{-14}T^3 [Pa\ s]$$
(99)

$$c_p = 1.00926 \times 10^3 - 4.0403 \times 10^{-2}T + 6.1759 \times 10^{-4}T^2 - 4.097 \times 10^{-7}T^3 \left[\frac{J}{kg\ K} \right]$$
(100)

in which the temperature T is in °C.

The porosity model used was proposed by [36], Eq. (101), valid for a circular-cylindrical geometry, which has been adapted to be used in an elliptic-cylindrical geometry considering $a = 1$ and $b = 20$. In this modified equation, the radial coordinate $\hat{r}$ represents the orthogonal distance of a given point within the reactor to the inner surface of the wall of the equipment.

In Eq. (101) J_0 is the Bessel function of first kind of zero order; $D_p = 0.004$ m is the particle diameter; $\hat{r}$ is the distance from any internal point to the reactor wall, which is normal to this wall, and $\varepsilon_o = 0.4$.

$$\varepsilon = \varepsilon_0 + (1 - \varepsilon_0)e^{-b\hat{r}}J_0\left(a\frac{\hat{r}}{D_p} \right)$$
(101)

5.2   Results and Analysis

The results presented here consist on the dimensionless temperature profiles $\left(\frac{T-T_m}{T_0-T_m} \right)$ in different planes: $z = 0.00556$, 0.0944 and 0.19444 m, $x = 0$ m, $y = 0$ m and $z = 0$ m and different process times, considering a mesh of $20 \times 20 \times 20$ control volumes and $\Delta t = 0.1$ s.

The temperature profile at the reactor inlet ($z = 0$ m) can be seen in Fig. 15, where the temperature varies from $T = 30$ °C at the reactor wall to $T = 95.44$ °C towards the center. The average temperature $T_M = 70$ °C in the cross section. In

Fig. 15 Temperature profile (°C) at the inlet of the elliptic-cylindrical reactor

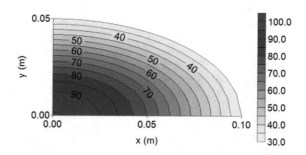

this new purpose, a fully developed thermal and hydrodynamic flow prior to reactor inlet was considered as a more realistic model for flow at the fluid inlet rather than the flat profile commonly used by many researchers. There are several experimental difficulties in dealing with temperature and velocity profiles in this critical region. In this area, the velocity profile tends to form preferential flow channels near the wall, which intensify the heat transfer to the reactor wall due to the formation of a wall-fluid contact thermal resistance [36, 37]. The formation tendency of these preferential channels affects the porosity profile in the fixed bed, which also presents much experimental difficulty for measurements.

In turn, the porosity model of this work, adapted from [36], agrees with the numerous experimental results reported in the literature [38–41], which are practically unanimous that the porosity of the bed presents unit value at a distance from which is of the order of half particle diameter, oscillating and decreasing to a constant value towards the center of the bed (Fig. 16). For a bed of spheres, this value is around $\varepsilon = 0.4$. Near the reactor wall, the maximum porosity favors the formation of the already mentioned preferential flow channels, which means that the fluid velocity profile at the inlet may be different from the piston profile, commonly considered by several authors.

Dong et al. [42] using a heterogeneous model for obtaining porosity via method of discrete elements, states that the porosity profile may not have radial symmetrical, because there may be differences in this parameter in different angular positions. The author affirms that two-dimensional pseudo-homogeneous models using a radial porosity profile may mask the radial asymmetry of the porosity profile, but

Fig. 16 Porosity profile inside the elliptic-cylindrical reactor in any axial position

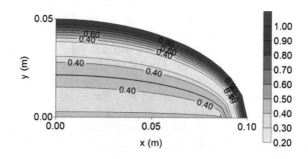

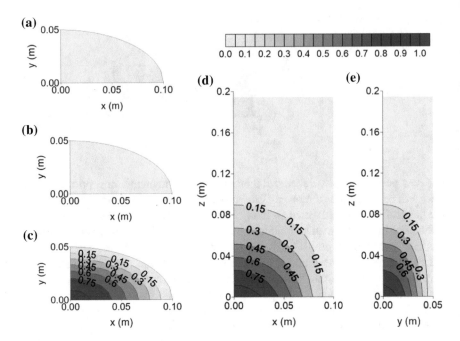

Fig. 17 Dimensionless temperature distribution inside the reactor at different planes (t = 0.5 s). **a** z = 0.1944 m, **b** z = 0.0944 m, **c** z = 0.0055 m, **d** y = 0 m and **e** x = 0 m

the use of a three-dimensional model helps to attenuate this deficiency presented by the two-dimensional model.

In relation to the transient effects on the reactor operation, it is possible to analyse relevant occurrences in the process.

Analysing Fig. 17a–e, it was found that after a time t = 0.5 s, the heated air stream percolating the reactor already provides rise in the axial and radial temperature gradients to z = 0.09 m. These gradients are still not as pronounced near the reactor wall because the temperature in that region is equal that of the refrigerant (thermal equilibrium). It is also observed at this time that there is virtually no disturbance in the reactor temperature profile from half the height towards the outlet (Fig. 17a, b).

At time t = 1.5 s, it can be seen in Fig. 18a–e that the higher axial temperature gradients are moving away from the reactor inlet region toward its outlet. Further, the radial and axial temperature gradients in the region after the value of z = 0.09 m, which practically do not exist at t = 0.5 s, now become very pronounced. It can be stated that up to time t = 1.5 s, a predominance of heat transfer in the axial direction by convection occurs.

At time t = 3 s, observing Fig. 19a–e and comparing with Fig. 17a–e, referring to time 1.5 s, it is found that part of the temperature profile is practically stabilized near the reactor inlet region—up to about a quarter of the height of the equipment—however, from half the height toward the outlet, the radial temperature gradients

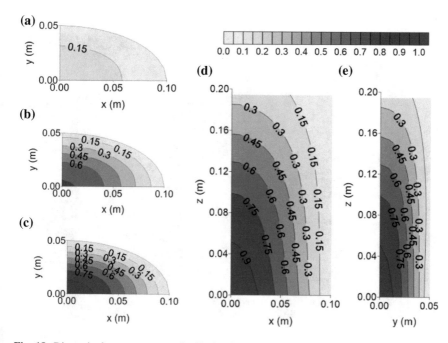

Fig. 18 Dimensionless temperature distribution inside the reactor at different planes (t = 1.5 s). **a** z = 0.1944 m, **b** z = 0.0944 m, **c** z = 0.0055 m, **d** y = 0 m and **e** x = 0 m

increase much more, especially in the region closest to the wall. It is also observed that the axial temperature gradients are becoming less pronounced along the bed height, indicating that, from that moment on, the higher heat transfer rate in the reactor has predominance in the radial direction.

The instant t = 4.5 s is the moment as the temperature inside the reactor reaches the steady state condition. By analysing Fig. 20a–e, it can be seen that the axial temperature gradients are quite small, especially on regions of the half of the half-axes towards the reactor wall. Furthermore, it is also observed that the radial temperature gradients at the end of the process increased significantly across the whole bed height, following the trend of the temperature profile of the fluid at the reactor inlet.

In general, the proposed model is highly versatile since that it is able to simulate different operating conditions involving a change in the geometry of the reactor; different materials and particle diameter; different velocity and temperature for the reactive and refrigerant fluids, being possible to simulate physical situations with different fluids, and with the consideration of the chemical reaction, in order to verify the system's temperature response.

With use of the pseudo-homogeneous model is not possible to simulate the behaviour of the liquid and solid phases separately, but its prognosis brings coherent physical information of the thermal behaviour of the reactor as operating.

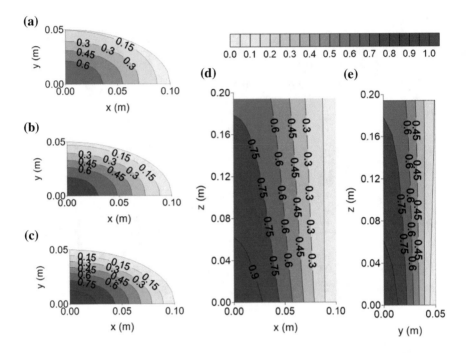

Fig. 19 Dimensionless temperature distribution inside the reactor at different planes (t = 3.0 s). **a** z = 0.1944 m, **b** z = 0.0944 m, **c** z = 0.0055 m, **d** y = 0 m and **e** x = 0 m

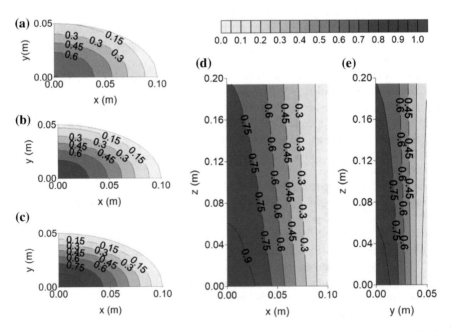

Fig. 20 Dimensionless temperature distribution inside the reactor at different planes (t = 4.5 s). **a** z = 0.1944 m, **b** z = 0.0944 m, **c** z = 0.0055 m, **d** y = 0 m and **e** x = 0 m

6 Concluding Remarks

In this chapter, the physical problem of conduction and convection heat transfer and fluid flow in porous media has been explored. Emphasis is given to packed bed reactor. Interest in this type of problem is motivated by its great importance in many practical applications of engineering and science.

Herein, an attention is focused on the unsteady-state heat transfer inside the reactor under local thermal equilibrium condition.

A well-posed transient and three-dimensional mathematical modelling for conduction/convection heat transfer within a packed-bed reactor of elliptic cross-section is proposed. Further, all general numerical formalism for the solution of the energy conservation equation, which is based on the finite-volume method, has been explored, and predicted results of temperature distribution inside the reactor at different instants are presented and discussed.

From the results, the following general conclusions can be summarized:

(a) The proposed model has the versatility to explore the transient simulation of several geometric forms of the reactor (circular-cylindrical, elliptic-cylindrical, including an approximation for rectangular channel), with various types of fluids, different particle and wall materials, with or without chemical reaction, as well as different boundary conditions at the fluid inlet and at the reactor surface;

(b) Transient modelling was physically consistent, incorporating the effects of parabolic profile of temperature at the reactor inlet on the temperature gradients. This procedure predominantly affected the radial gradients, with consequent higher heat transfer rate in this direction;

(c) The proposed porosity model, although adapted from a model with circular cylindrical geometry with low tube diameter/particle diameter ratio, proved to be quite satisfactory for the elliptic-cylindrical shape, preserving the expected physical aspects;

(d) The largest radial temperature gradients occur in the direction of the y-axis—the minor half-axis of the ellipse—standing out the extremity of this region. For any time and height in this region occurs higher heat transfer and higher surface temperature;

(e) At the beginning of the process, the largest axial gradients are at the reactor inlet and, over time, are reduced in this region and intensify towards the exit.

As a final comment we start that models, such as outlined in this chapter, can be used with great confidence to elucidate unknown features of several multiphase complex systems, mainly related to porous media and packed bed reactors.

Acknowledgements The authors thank to FINEP, CAPES and CNPq (Brazilian Research Agencies) for financial support to this research, and also to the researchers for their referenced studies which helped in improving the quality of this work.

References

1. Dullien, F.A.L.: Porous Media—Fluid Transport and Pore Structure, 2nd edn, p. 574. Academic Press Inc, San Diego, USA (1992)
2. Martins, A.A.A.: Transport phenomena in porous media. Single phase flow and mass transport. Ph.D. Thesis, Faculty of Engineering (FEUP), University of Porto, Porto, Portugal (2006) (in Portuguese)
3. McCabe, W.L., Smith, J.C., Harriot, P.: Unit Operations of Chemical Engineering. Chemical Engineering Series, 5th edn, p. 1129. McGraw Hill, USA (1993)
4. Rawlings, J.B., Ekerdt, J.G.: Chemical Reactor Analysis and Design Fundamentals, p. 609. Nob Hill Publishing, Madison, WI, USA (2002)
5. Froment, G.F., Bischoff, K.B., De Wilde, J.: Chemical Reactor Analysis and Design, 3rd edn, p. 859. Wiley, NewYork, USA (2011)
6. Eppinger, T., Seidler, K., Kraume, M.: DEM-CFD simulations of fixed bed reactors with small tube to particle diameter ratios. Chem. Eng. J. **166**(01), 324–331 (2011)
7. Hill Jr., C.G.: An Introduction to Chemical Engineering Kinetics and Reactor Design, p. 594. Wiley, NewYork (1977)
8. Oliveira, L.G.: Heat transfer in packed bed cylindrical-elliptical reactor: thermal, fluiddynamics and geometric aspects. Ph.D Thesis in Process Engineering, Federal University of Campina Grande, Campina Grande, Brazil, (2004) (in Portuguese)
9. Pirkle Jr., J.C., Reyes, S.C., Hagan, P.S., Khesgid, H.: Solution of dynamic distributed parameter model of nonadiabatic fixed—bed reactor. Comput. Chem. Eng. **11**(06), 737–747 (1987)
10. Shafeeyan, M.S., Daud, W.M.A.W., Shamiri, A.: A review of mathematical modeling of fixed-bed columns for carbon dioxide adsorption. Chem. Eng. Res. Des. **92**(5), 961–988 (2014)
11. Jee, J.G., Park, H.J., Haam, S.J., Lee, C.H.: Effects of nonisobaric and isobaric steps on O_2 pressure swing adsorption for a reactor. Ind. Eng. Chem. Res. **41**(17), 4383–4392 (2002)
12. Kim, M.B., Bae, Y.S., Choi, D.K., Lee, C.H.: Kinetic separation of landfill gas by a two-bed pressure swing adsorption process packed with carbon molecular sieve: non isothermal operation. Ind. Eng. Chem. Res. **45**(14), 5050–5058 (2006)
13. Kim, M.B., Moon, J.H., Lee, C.H., Ahn, H., Cho, W.: Effect of heat transfer on the transient dynamics of temperature swing adsorption process. Korean J. Chem. Eng. **21**(3), 703–711 (2004)
14. Amiri, A., Vafai, K.: Transient analysis of incompressible flow through a packed bed. Int. J. Heat Mass Transf. **41**, 4259–4279 (1998)
15. Ergun, S.: Fluid flow through packed columns. J. Chem. Eng. Prog. **48**(2), 89–94 (1952)
16. Benanati, R.F., Brosilow, C.B.: Void fraction distribution in beds of spheres. Am. Inst. Chem. Eng. J. **8**(3), 359–361 (1962)
17. Wakao, N., Kaguei, S.: Heat and Mass Transfer in Packed Beds. Gordon and Breach Science Publishers, New York (1982)
18. Colburn, A.P.: Heat transfer and pressure drop in empty, baffled and packed tubes: I. Heat transfer in packed tubes. Ind. Eng. Chem. Res. **23**(8), 910–913 (1931)
19. de Azevedo, S.F., Romero, M.A.O., Wardle, A.P.: Modeling of tubular fixed-bed catalytic reactor: a brief review. Chem. Eng. Res. Des. **68**, 483–502 (1990)
20. Moreira, M.F.P., Ferreira, M.C., Freire, J.T.: Evaluation of pseudo-homogeneous models for heat transfer in packed beds with gas flow and gas-liquid cocurrent downflow and upflow. Chem. Eng. Sci. **61**(06), 2056–2068 (2006)
21. Coberly, C.A., Marshall Jr., M.W.R.: Temperature gradients in gas streams flowing through fixed granular beds. Chem. Eng. Prog. **47**(3), 141–150 (1951)
22. Dixon, A.G., Paterson, W.R., Cresswell, D.L.: Heat transfer in packed beds of low tube/particle diameter ratio. ACS Symp. Ser. **65**, 238–253 (1978)

23. Borkink, J.G.H., Westerterp, K.R.: Determination of effective heat transport coefficients for wall-cooled packed beds. Chem. Eng. Sci. **47**(9–11), 2337–2342 (1992)
24. Giudici, R.: Modeling of ethylene oxidation reactor: study of thermal parameters and strategy of catalyst dilution. Ph.D. Thesis, Poli/USP, São Paulo, Brazil, p. 183(1990) (in Portuguese)
25. Silva, R.M., Lima, A.G.L., Oliveira, L.G., Araújo, M.V., Santos, R.S.: Transient heat transfer in a packed-bed elliptic cylindrical reactor: a finite-volume approach. Defect and Diffusion Forum **380**, 79–85 (2017)
26. Lima, A.G.B.: Diffusion phenomena in prolate spheroidal solids. Studied case: drying of Banana. Ph.D Thesis in Mechanical Engineering, State University of Campinas, Campinas, Brazil, p. 244 (1999) (in Portuguese)
27. Kreyszig, E.: Advanced engineering Mathematics, vol. 1, 10th edn. Wiley, New York (2011)
28. Maliska, C.R.: Heat transfer and computational fluid mechanics. 2nd edn, LTC—Livros Técnicos e Científicos Editora S.A., Rio de Janeiro, Brazil (2004) (in Portuguese)
29. Magnus, W., Oberhettinger, F., Soni, R.P.: Formulas and Theorems for the Special Functions of Mathematical Physics, 3rd edn. Springer, Berlin (1966)
30. Brodkey, R.S.: The Phenomena of Fluid Motions. Addison-Wesley Publishing Company, London (1967)
31. Abramowitz, M., Stegun, I.: Handbook of Mathematical Functions with Formulas, Graphs, and Mathematical Tables. Wiley, New York (1970)
32. Bergman, T.L., Lavine, A.S., Incropera, F.P., De Witt, D.P.: Fundamentals of Heat and Mass Transfer, 7th edn, LTC—Livros Técnicos e Científicos Editora S.A, Rio de Janeiro, Brazil (2014). (in Portuguese)
33. Patankar, S.V.: Numerical Heat Transfer and Fluid Flow. Hemisphere Publishing Corporation, New York, USA (1980)
34. Pakowski, Z., Bartczak, Z., Strumillo, C., Stenström, S.: Evaluation of equations approximating thermodynamic and transport properties of water, steam and air for use in cad of drying processes. Drying Technol. **09**(3), 753–773 (1991)
35. Jumah, R.Y., Mujumdar, A.S., Raghavan, G.S.V.: A Mathematical model for constant and intermittent batch drying of grains in a novel rotating jet spouted bed. Drying Technol. **14**(03–04), 765–802 (1996)
36. Taylor, K., Smith, A.G., Ross, S.: The prediction of pressure drop and flow distribution in packed-bed filters. In: Second International Conference on CFD in the Minerals and Process Industries (CSIRO), Melbourne, Australia, 273 (1999)
37. Schwartz, C.E., Smith, J.M.: Flow distribution in packed beds. Ind. Eng. Chem. Res. **45**(6), 1209–1218 (1953)
38. Zotin, F.M.Z.: The wall effect on filling columns, Masters Dissertation in Chemical Engineering, Federal University of São Carlos, São Carlos, Brazil (1985)
39. Kufner, R., Hofmann, H.: Implementation of radial porosity and velocity distribution in a reactor model for heterogeneous catalytic gas phase reactions (TORUS-model). Chem. Eng. Sci. **45**(8), 2141–2146 (1990)
40. Mueller, G.E.: Radial void fraction distributions in randomly packed-fixed beds of uniformly sized spheres in cylindrical containers. Powder Technol. **72**(3), 269–275 (1992)
41. de Klerk, A.: Voidage variation in packed beds at small column to particle diameter ratio. Am. Inst. Chem. Eng. J. **49**(8), 2022–2029 (2003)
42. Dong, Y., Sosna, B., Korup, O., Rosowski, F., Horn, R.: Investigation of radial heat transfer in a fixed-bed reactor: CFD simulations and profile measurements. Chem. Eng. J. **317**, 204–214 (2017)

Advanced Study to Heat and Mass Transfer in Arbitrary Shape Porous Materials: Foundations, Phenomenological Lumped Modeling and Applications

E. S. Lima, W. M. P. B. Lima, Antonio Gilson Barbosa de Lima, S. R. de Farias Neto, E. G. Silva and V. A. B. Oliveira

Abstract This chapter provides information related to simultaneous heat and mass transfer in unsaturated porous bodies with particular reference to drying process of arbitrarily-shaped solids. Several important topics such as drying theory, moisture migration mechanisms, lumped and distributed modeling for homogeneous and heterogeneous bodies, and applications are presented and discussed. Herein, a new phenomenological and advanced lumped-parameter model written in any coordinate system is presented, and the analytical solutions of the governing equations, limitations of the modeling and general theoretical results are discussed. The

E. S. Lima
Department of Mathematics, State University of Paraiba, R. das Baraúnas,
S/N, Campina Grande, PB 58429-500, Brazil
e-mail: limaelisianelima@hotmail.com

W. M. P. B. Lima · A. G. Barbosa de Lima (✉)
Department of Mechanical Engineering, Federal University of Campina Grande,
Av. Aprígio Veloso, 882, Bodocongó, Campina Grande, PB 58429-900, Brazil
e-mail: antonio.gilson@ufcg.edu.br

W. M. P. B. Lima
e-mail: wan_magno@hotmail.com

S. R. de Farias Neto
Department of Chemical Engineering, Federal University of Campina Grande,
Av. Aprígio Veloso, 882, Bodocongó, Campina Grande, PB 58429-900, Brazil
e-mail: fariasn@gmail.com

E. G. Silva
Department of Physics, State University of Paraiba, R. das Baraúnas,
351, Campina Grande, PB 58429-500, Brazil
e-mail: dinasansil@yahoo.com.br

V. A. B. Oliveira
State University of Paraiba, Rodovia PB 075, S/N, km 1, Guarabira,
PB 58200-000, Brazil
e-mail: profvitaloliveira@gmail.com

J. M. P. Q. Delgado and A. G. Barbosa de Lima (eds.), *Transport Phenomena
in Multiphase Systems*, Advanced Structured Materials 93,
https://doi.org/10.1007/978-3-319-91062-8_6

proposed model includes different effects such as shape of the body (hollow or not hollow), heat and mass generation, and coupled heating, evaporation and convection phenomena.

Keywords Drying · Heat · Mass · Theoretical · Complex shape solid

1 Drying Theory of Wet Porous Solids

1.1 Foundations

Drying is a complex process involving simultaneous phenomena of heat, and mass transfer and linear momentum, existence of equilibrium state and dimensional variations of the solid being dried. Drying differs from other separation techniques, such as osmotic dehydration, by the way like water is removed from the solid. In the drying the removal of water molecules occurs by movement of the liquid, due to a difference of partial pressure of the water vapor between the surface of the product and the air that surrounds it.

Heat can be supplied to the material to be dried by: thermal radiation, convection, conduction or by volumetric absorption of the electromagnetic energy generated at the radio or microwave frequency. This volumetric heat transfer can accelerate the drying process and offers a number of benefits over conventional methods. In most cases, heat transfer occurs through a combination of several mechanisms.

The products are very different from each other, due to their composition, structure, shape and dimensions. The drying conditions are very diverse, according to the drying air properties and the form like the air-product contact occur. As examples, we can cite drying of wet particles bed with hot air and suspension of particles in an air stream. Once the product is placed in contact with hot air, there is heat transfer from the air to the product due to the temperature difference between them. Simultaneously, the difference in the partial pressure of water vapor between the air and the surface of the product determines a transfer of matter (mass) from the product surface to the air. The latter is make in the form of water vapor. We start that part of the heat that reaches the product is used to water evaporation. Figure 1 illustrates the curves of the water content of the product, its temperature and the drying rate, over time, when air is used as a drying agent.

The curve (a) represents the reduction of the moisture content of the product during drying. It represents the mass of water per mass of dry product ratio during the process in a drying condition previously established. The curve (b) represents the product drying velocity (drying rate) which represent the moisture content variation of the product as a function of time, being obtained from differentiating the curve (a) with respect to time. The curve (c) represents the variation of product

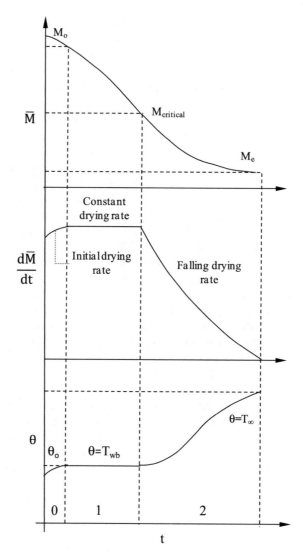

Fig. 1 Moisture content and temperature transient behavior of the product during drying

temperature during drying. It is obtained by measuring the temperature of the product during the process.

During drying, the product goes through several stages, depending on its initial moisture content, nature and form. The following describes each one of them (Fig. 1):

(a) **Period 0**: This is the period of induction or the period to enter in operating regime (warming up or accommodation period of the material). In the beginning, the product is usually at a temperature lower than the air, and the partial pressure of water vapor at the surface of the product is low, consequently, the mass transfer and drying rate are also low. The heat reaching the material in large quantities causes an increase in temperature, causing an increase in partial pressure of water vapor in the air and in the drying rate. This phenomenon continues until the heat transfer to compensate exactly for mass transfer. If the air temperature is lower than the product, so, the product temperature will decrease until it reaches the thermal equilibrium. The duration of this period is insignificant relative to the total drying time.

(b) **Period 1**: This period corresponds to the constant drying velocity (drying rate). During this period, as in the previous one, the amount of water available within the product is quite large. Water evaporates as free water at the surface of the product. The partial pressure of water vapor at the surface is constant and equal to the vapor pressure of pure water at the temperature of the product. The product temperature, at this time is also constant and equal to the wet bulb temperature of the drying air, because the compensation between heat and mass transfers exactly. The drying velocity is therefore constant. This period continues, while the migration of water from the inside to the surface of the product is enough to keep up with the evaporative loss of water at the surface. In the case of biological materials is very difficult the existence of this period, because the drying operating conditions provokes a mass transfer resistance essentially inside the product, causing the water evaporation rate from the surface to the environment higher than the replacement rate of moisture from the interior to the surface of the material.

(c) **Period 2**: This is the period of falling drying rate. It starts as soon as the water begins to reach the surface in quantities less than the water evaporated, so, the drying velocity decreases. Some authors define the value of moisture content of the product at the transition point between periods 1 and 2 as being the critical moisture content, however, we starts that this is not a physical property of the material. The critical moisture content depends on the drying operating conditions and solid geometry. During this period, the heat exchange is no longer compensated for the loss in mass, therefore, the temperature of the product increases and tends asymptotically to the air temperature (thermal equilibrium). Throughout this period the limiting factor is the internal migration of water. This reduction in the drying rate is sometimes interpreted by decreasing the wet surface, but the most frequent interpretation is by lowering the partial pressure of water vapor at the surface. At the end of this period the product will be in equilibrium with the air and the drying velocity is null (hygroscopic equilibrium).

In the drying process there are some factors that can interfere in the attributes of quality, among the main ones can be highlighted those that are directly linked to the product to be dried such as thickness and characteristics of the product, and also, with those that are connected with drying air (temperature, velocity and relative humidity). The drying process occurs in the natural form exposing the product directly to the sun, and artificially using equipment such as oven and dryers [1].

The mechanisms of internal mass transfer during drying of biological materials, can be influenced by two collateral phenomena during drying, cited below:

(a) Existence of the solute's contribution during drying. For example, plum sugar (solute) is deposited at the surface during drying, forming a crust that decreases the drying velocity. Another example is the drying of beet. This product dries more quickly when its sugar content is reduced before drying.

(b) Biological products are living cells thus exhibiting a specific behavior where the cell is stretched by the liquid contained therein and, as a consequence, the cell wall is subjected to stress and the liquid contained therein is subjected to compression. This phenomenon is known as "turgor". As the drying proceeds, with the removal of water, there is reduction in the pressure that the liquid exerts in the cell wall. The phenomena associated with this pressure reduction are treated as a result of the shrinkage of the material. During the drying process, the shrinkage phenomenon provokes several effects in the material. As the material shrinks while during drying, the surface of the material hardens ("case hardening"), thus, the material deform and fissures, especially in materials very sensitive to heat.

1.2 Moisture Migration Mechanisms

The phenomenon of moisture migration inside of porous materials is still not well known. Some authors claim that migration can be a combination of moisture movement by diffusion of liquid and vapor, each predominating in certain stages of drying [2]. Wherefore, several theories have been proposed to describe the mass and heat transport in capillary porous media, such as: Liquid diffusion theory; Capillary theory; Krischer's theory; Luikov's theory; Philip and De Vries theory; Berger and Pei theory, and the Fortes and Okos theory. A detailed discussion of drying theories can be found in the literature [3–9].

According to theories listed above, the following moisture transport mechanisms in solids have been provided by literature [3, 7, 9–11]:

(a) transport by liquid diffusion due to moisture concentration gradients;
(b) transport by diffusion of vapor due to moisture concentration gradients and vapor partial pressure (caused by temperature gradients);
(c) transport by effusion (Knudsen flow). This occurs when the mean free path of the molecules of vapor is the same order of magnitude of pore diameter. It is important for high vacuum conditions, like for example, lyophilization;
(d) transport of vapor by thermodiffusion due to temperature gradients;
(e) transport of liquid by capillary force due to capillary phenomena;
(f) transport of liquid by osmotic pressure due to osmotic force;
(g) transport of liquid due to gravity;

(h) transport of liquids and vapor, due to the total pressure difference, caused by external pressure, contraction, high temperature and capillarity;

(i) transport by surface diffusion due to migration of liquid and vapor mixture though the pores of the product surface.

Additional information of each moisture transport mechanisms listed above can be found in the works cited in this item.

1.3   Types of Drying

The drying process can be accomplished in several ways in order to several purposes. For food products, for example, is employed mainly in conservation, while also allowing the transportation and storage without refrigeration. The most common drying types are:

(a) **Stationary drying**: The designation of stationary drying is given when there is no movement of the product during drying. It is generally used in grains, which are placed in silos-dryers that undergo the action of heated air. This type of drying presents low performance as a function of the height of the solid layer which is regulated by the distance of the drying entrance and the air flow rate involved.

(b) **Continuous drying**: In this drying system, solids flows in the dryer while are heated. This process allows the loss of water from the product and its heating. in such a way that they enter wet in the dryer, receives heating action, and leaving it dried. In this type of drying, the system can be classified as follows:
Concurrent flow: the product along with the drying air moves in parallel, in the same directions inside the equipment;
Countercurrent flow: the product together with the drying air travels parallel in opposite directions inside the equipment;
Cross flow: the drying air moves perpendicularly through the mass of the product.
Mixed flow: This is a combination of different drying types.

(c) **Intermittent drying**: Intermittent drying and a type of discontinuous drying, with periods of energy and heat application. It is characterized by the discontinuous passage of air by the mass of the product in movement, promoted by the recirculation of the solid in the dryer. Intermittent drying controls the rate of heat input for the material to be dried, in order to avoid thermal degradation of heat-sensitive products.

2 Drying Process Modeling

One of the most important drying technologies, especially for industrial processes, is the mathematical modeling of processes and drying equipment. One purpose of modeling is to enable the engineer to choose the most appropriate drying method of a specific product as well as suitable operating conditions. The principle of modeling is based in a mathematical equation system that characterizes completely the physical system being modeled. In particular, the solution of the governing equations makes it possible to predict the process parameters as a function of the drying time only based on the initial and boundary conditions and simplification, and some required data of the product and air. The starting point in mathematical modeling is the definition of the modification process, in particular a description of the input data that influence the process, as well as variables that depend on the behavior of the process.

From the physical point of view, the drying process involves simultaneous phenomena of heat and mass transfer, linear momentum and dimensional variations of the product. It is a very complex process involving several physical phenomena, and thus, there is a need to generate mathematical models to simulate the drying process with great physical realism. Then, it is important to insert in the drying model the maximum amount of information related to the process, allowing to correctly relate the model to the actual physical situation. Thus, the development of mathematical models to describe drying process has been subject of study for many researchers for decades.

Recently many sophisticated drying models have been presented in the literature. Depending on the layer thickness of the studied material, these models can be classified in thin layer (particle level models) and thick layer models (dryer level models). The practical importance of thin-layer drying has limitations, because the materials are usually dried in thick layers: stationary or in motion. However, the models most used by the researchers take into account thermo physical properties, drying kinetics, and mass and energy balance in the dryer, thus ratifying the need to have an equation for describe thin layer drying kinetics of the material under certain pre-established operating conditions. Therefore, numerous thin layer models have been proposed to describe the moisture loss rate during drying of solids and can be divided into two main groups: lumped models and distributed models, which will be detailed below.

2.1 Distributed Models

The distributed models describe the heat and mass transfer rates as a function of the position within the wet porous solid and the drying time. They consider the external and internal resistances to heat and mass inside the porous solids. Thus, these models or systems are based on the interaction between time, and one or more spatial variables for all their dependent variables. In this approach, the gradients of moisture content, pressure and temperature inside the material are considered [12]. Some of these models reported in the literature are described below.

(a) Luikov's model

This model is based on the thermodynamics of the irreversible processes and proposes that the water moves in porous capillary media, in isothermal conditions, due to the action of a potential gradient of mass transfer. This potential of mass transfer was created by Luikov based on the analogy with the driving force of heat transfer, the temperature gradient [4].

Luikov [13] presented a mathematical model to describe the drying process of porous capillary products based on the mechanisms of diffusion, effusion, vapor convection and diffusion and convection of water inside the porous medium. The process is described by a system of partial differential equations coupled for temperature, moisture content and in cases of intense drying also the pressure. The set of equations is as follows:

$$\frac{\partial M}{\partial t} = \nabla^2(K_{11}M) + \nabla^2(K_{12}\theta) + \nabla^2(K_{13}P) \tag{1}$$

$$\frac{\partial \theta}{\partial t} = \nabla^2(K_{21}M) + \nabla^2(K_{22}\theta) + \nabla^2(K_{23}P) \tag{2}$$

$$\frac{\partial P}{\partial t} = \nabla^2(K_{31}M) + \nabla^2(K_{32}\theta) + \nabla^2(K_{33}P) \tag{3}$$

where K_{ij}, i, j = 1, 2 and 3, are the phenomenological coefficients for i = j and the combined coefficients for i $\neq$ j.

(b) Diffusive models

Several authors consider the diffusion of liquid water as the main transport mechanism of moisture in wet porous solids [1, 11, 14–23]. To describe the drying process theoretically the Fick's second law has been widely used since it establishes the diffusion of moisture in terms of the concentration gradient in the solid, as follows:

$$\frac{\partial M}{\partial t} = \nabla \cdot (D\nabla M) \tag{4}$$

In general, the diffusion coefficient D, is considered constant, or dependent on the temperature and/or the moisture content of the solid. However, it is worth noting that mechanical compression reduces porosity and effective moisture diffusivity; therefore, the pressure has a negative effect on water diffusivity [24]. Table 1 provides a summary of some of the various empirical parametric models expressing moisture diffusivity as a function of temperature and/or moisture content reported in the literature.

Table 1 Empirical parameters models for mass diffusion coefficient [15]

Parametric model	Equation number
$D(M,T) = A_0 \exp(A_1 M) \exp\left(-\dfrac{A_2}{T_{abs}}\right)$	(5)
$D(M,T) = A_0 \exp\left(-\dfrac{A_1}{M}\right) \exp\left(-\dfrac{A_2}{T_{abs}}\right)$	(6)
$D(M,T) = A_0 \exp\left(\sum_{i=1}^{3} A_i M^i\right) \exp\left(-\dfrac{A_4}{T_{abs}}\right)$	(7)
$D(M,T) = A_0[1 - \exp(-A_1 M)] \exp\left(-\dfrac{A_2}{T_{abs}}\right)$	(8)
$D(M,T) = A_0[1 + \exp(A_1 - A_2 M)]^{-1} \exp\left(-\dfrac{A_3}{T_{abs}}\right)$	(9)
$D(M,T) = A_0 \exp(A_1 M) \exp\left(-\dfrac{A_2 M + A_3}{T_{abs}}\right)$	(10)
$D(M,T) = A_0(M) \exp\left[-\dfrac{A_1 \exp(-A_2 M) + A_3}{T_{abs}}\right]$	(11)
$D(M) = (A_0 + A_1 M)$	(12)
$D(T) = A_0 \exp\left(-\dfrac{A_1}{R_0 T_{abs}}\right)$	(13)

The concept of liquid diffusion as the only mechanism of moisture transport has been the subject of several criticisms, constantly presenting discrepancies between experimental and theoretical values [4, 6]. The Luikov's model, considering negligible the effects of pressure gradients and combined temperature and moisture content, is similar to the diffusion model.

(c) Model based on the non-equilibrium thermodynamics

Based on the thermodynamic concepts of irreversible processes, Fortes and Okos [25] have proposed a model that considers the existence of local equilibrium valid; the use of the Gibbs equation for non-equilibrium conditions; of the phenomenon of shrinkage negligible; neglected total pressure effects; the use of linear phenomenological laws; validity of the fundamental relations of Onsager; of the system be taken as isotropic; of the water to migrate in the liquid and vapor phases; of being the rate of heat transfer and mass slower than the rate of phase change, and finally, of the use of the Curie principle.

According to Fortes and Okos [25], the fundamental difference between the theory of these authors and the theories cited is that the driving force for the isothermal movement of both liquid and vapor is a gradient of the equilibrium moisture content that the product attains when subjected for a sufficiently long time to controlled conditions of temperature and relative humidity of the air, and not the moisture content gradient. The driving force for the liquid and vapor transfer is the chemical potential gradient, which in turn is a function of the temperature, relative humidity and equilibrium moisture content. In this model, it is postulated that water in porous capillary media can moves in the opposite direction to the moisture content gradient, but always in the direction of the equilibrium moisture content gradient. Thus, equilibrium moisture content is presented as a more natural choice for mass transport potential than the concept proposed by Luikov.

Admitting the considerations of the model, considering the negligible effects of gravity on the vapor transport and using Onsager relations, Fortes and Okos [25] derived the following equations, for capillary porous bodies:

- **Heat flux**

$$\vec{J}_q = -k_t \nabla T - \left[\rho_l k_l R_v \ln(UR) + k_v \left(\rho_{vo} \frac{\partial UR}{\partial T} + UR \frac{d\rho_{vo}}{dT} \right) \right]$$
$$\times \frac{R_v T^2}{UR} \frac{\partial UR}{\partial T} \nabla M + T \left[\rho_l k_l R_v \ln(UR) + k_v \left(\rho_{vo} \frac{\partial UR}{\partial T} + UR \frac{d\rho_{vo}}{dT} \right) \right] \vec{g}$$

$$(14)$$

- **Liquid flux**

$$\vec{J}_l = -\rho_l k_l R_v \ln(UR) \nabla T - \rho_l k_l \frac{R_v T}{UR} \left(\frac{\partial UR}{\partial M} \right) \nabla M + \rho_l k_l \vec{g} \qquad (15)$$

- **Vapor flux**

$$\vec{J}_v = -k_v \left(\rho_{vo} \frac{\partial UR}{\partial T} + UR \frac{d\rho_{vo}}{dT} \right) \nabla T - k_v \rho_{vo} \left(\frac{\partial UR}{\partial M} \right) \nabla M \qquad (16)$$

Assuming that no ice is present and that the air mass is negligible, one can write the mass conservation equation as

$$\frac{\partial (\rho_{ps} M)}{\partial t} = -\nabla \cdot \left(\vec{J}_l + \vec{J}_v \right) \qquad (17)$$

With the consideration of non-existence of shrinkage phenomena, ρ_{ps} constant, the Eq. (17) reduces to:

$$\rho_{ps} \frac{\partial M}{\partial t} = -\nabla \cdot \left(\vec{J}_l + \vec{J}_v \right) \qquad (18)$$

The energy conservation equation can be obtained, assuming that the rate of change of enthalpy of the system minus the heat of adsorption equal to the divergence of the enthalpy flux. Thus, one can write:

$$\rho_{ps}c_p \frac{\partial T}{\partial t} - \rho_{ps}h_w \frac{\partial M}{\partial t} = -\nabla . \overrightarrow{J}_q - h_{fg}\nabla . \overrightarrow{J}_v - \overrightarrow{J}_1 . c_l\nabla T - \overrightarrow{J}_v . c_v\nabla T \quad (19)$$

As can be seen, this model more accurately describes the physics of the heat and mass transfer process than the simple liquid diffusion model, however its applicability is greatly limited, in virtue of the ruling equations of the phenomenon include many coefficients that are difficult to determine experimentally, depending on the product nature.

2.2 Lumped Models

The thin-layer drying equations, namely lumped model, can be classified as empirical, semi-empirical or semi-theoretical, and purely theoretical. This classification is given based on their comparative advantages and disadvantages and also its derivation [26–30]. These equations neglect the effects of temperature and moisture content gradients inside the material during the drying process. Some of these models assume that the material reaches the drying air temperature immediately at the beginning of the process.

Empirical equations have a direct relationship between moisture content and drying time, while the semi-empirical equations are analogous to Newton's law of cooling, assuming the drying rate is proportional to the difference between the moisture content of the product and its respective equilibrium moisture content for a specified drying conditions. The semi-theoretical equations are usually obtained from the liquid and/or vapor diffusion equation within the product.

In short, the lumped models describe the heat and mass transfer rates for the entire product, ignoring internal resistance of heat and mass transfer. Many lumped equations are derived from the distributed equations under small considerations. The thin-layer drying models are often employed to describe the drying of fruits and vegetables; however, because of their ease and fast response of the problem, it has been used to describe the drying of non-biological materials also. These categories of models take into consideration only external resistance to the moisture transport process between the material and atmospheric air, providing a greater extent of accurate results of the drying process, and make less assumption due to its dependence with experimental data. Thus, these models proved to be the most useful for engineers and designers of dryers [11], but, are only valid in the drying conditions applied. However, among them, the theoretical models make several assumptions leading to considerable errors [31–34], thereby limiting the use in the design of dryers.

The main challenges faced by empirical models are that they depend on largely of experimental data and provide limited information about the heat and mass transfer during the drying process [12, 35]. Due to the characteristics of the semi-theoretical and empirical models, these models are widely applied in the estimation of drying kinetics of products with high moisture content such as fruits and vegetables. Following are detailed some of these lumped models reported in the literature.

2.2.1 Semi-theoretical Models

The semi-theoretical models are derived from the distributed model (Fick's second law of diffusion) or its simplified variation (Newton's law of cooling). Semi-theoretical models of Lewis, Page and Page Modified are derived from Newton's law of cooling, while to following models are derived from Fick's second law of diffusion [12]:

(a) Exponential model and simplified form;
(b) Exponential model of 2-terms and modified form.
(c) Exponential model of 3-terms and simplified form.

Factors that could determine the application of these models for a specific product include the drying condition (relative humidity, temperature and air velocity of the air), shape, dimension, and the initial moisture content of the material to be dried [9, 12, 27]. In addition, under these conditions may be noted that the complexity of the models can be attributed to the number of constants that appears in them. How much large the number of parameters to be determined more complex will be the model. Then, under this mathematical point of view the Newton model is the simplest. However, the selection of the most appropriate model to describe the behavior of drying of a particular product does not depend exclusively on the number of constants, but also on the result of several statistical indicators reported in the literature [12, 29, 36–50]. Following will be detailed some semi-theoretical models reported in the literature.

(a) **Models derived from Newton's law of cooling**.

Newton's Model. This model is sometimes cited in the literature as the Lewis model, exponential model or single exponential model. It is said that it is the simplest model because it contain only one constant to be determined. In the past, this model was widely applied in describing the drying behavior of various foods and agricultural products; however, currently it has been occasionally usage to describe the drying behavior of some fruits and vegetables. The following equation represent the Newton's model:

$$\overline{M}^* = \frac{\overline{M} - M_e}{M_0 - M_e} = \exp(-K_1 t) \tag{20}$$

In Eq. (20), k is the drying constant (s^{-1}). $\overline{M}^*$ is the dimensionless moisture content, M is the moisture content on a dry basis at any time t. M_0 is the initial moisture content in of the sample base dry and M_e is the equilibrium moisture content.

Page's Model. The Page model or the modified Lewis model is an empirical modification of the Newton's model, in which errors associated with the use of the Newton's model are greatly reduced by the addition of a dimensionless empirical constant (n) in the time, as follows:

$$\overline{M}^* = \frac{\overline{M} - M_e}{M_0 - M_e} = \exp(-K_1 t^n) \qquad (21)$$

Modified Page's Model. This model is a modified form of the Page's model. The following equations represents this cited model:

$$\overline{M}^* = \frac{\overline{M} - M_e}{M_0 - M_e} = a_1 \exp\left[-\left(\frac{t}{K_1^2}\right)^n\right] \qquad (22)$$

where a_1 and n are empirical constants (dimensionless). Equation (22) is a variant of the Page's model, however others variation are reported in the literature.

(b) **Models derived from Fick's second law of diffusion.**

Henderson and Pabis model. This model is the first term of the general solution of the Fick's second law of diffusion. It can also be considered as a simple model with only two constants. The equation of this model is given below:

$$\overline{M}^* = \frac{\overline{M} - M_e}{M_0 - M_e} = a_1 \exp(-K_1 t) \qquad (23)$$

where a1 and K_1are constants to be determined.

Modified Henderson and Pabis model. The modified Henderson and Pabis model is a general solution with 3 terms of Fick's second law of diffusion for the correction of the deficiencies that occurs as using the Henderson and Pabis model. It has been reported that the first term explains the last part of the drying process of food and agricultural products, which occurs largely in the last period of falling rate, the second term describes the intermediate part and the third term explains the loss of initial moisture in the drying [12]. The model contains 6 constants, and thus this model was called complex thin-layer model.

However, it should also be emphasized that with 6 parameters, many more than 6 experimental data points are required to calculate the model. The model is not so complex with the advent of computers, but, statistically speaking a good degree of freedom is necessary for reliability of results, and this will require a lot of experimental data.

$$\overline{M}^* = \frac{\overline{M} - M_e}{M_0 - M_e} = a_1 \exp(-K_1 t) + a_2 \exp(-K_2 t) + a_3 \exp(-k_3 t) \quad (24)$$

where a_i are dimensionless model constants, and k_i are the drying constants, which, must be determined by experimental data.

Midilli's model. Midilli et al. [51] proposed a new model through a modification in the Henderson and Pabis model by adding an extra term with a coefficient. The new model, which is a combination of an exponential term and a linear term, is given as follows:

$$\overline{M}^* = \frac{\overline{M} - M_e}{M_0 - M_e} = a_1 \exp(-K_1 t) + a_2 t \quad (25)$$

where a_1 and a_2 are the model constants and K_1 is the drying constant to be estimated from the experimental data.

Logarithmic model. This model is also known as an asymptotic model and is another modified form of the Henderson and Pabis model. It is actually a logarithmic form of the Henderson and Pabis model with the addition of an empirical term. The model contains 3 constants and can be expressed as follows:

$$\overline{M}^* = \frac{\overline{M} - M_e}{M_0 - M_e} = a_1 \exp(-K_1 t) + a_2 \quad (26)$$

where a_1 and a_2 are empirical constant without dimension.

Two-term model. According to Sacilik [52], the 2-term model is the general solution with 2 terms of Fick's second law of diffusion. The model contains 2 dimensionless empirical constants and 2 model constants that can be determined from experimental data. The first term describes the last part of the drying process, while the second term describes the beginning of the drying process. This model well describes the moisture transfer in drying process. The equation related to this model is given below:

$$\overline{M}^* = \frac{\overline{M} - M_e}{M_0 - M_e} = a_1 \exp(-K_1 t) + a_2 \exp(-K_2 t) \quad (27)$$

where a_1 and a_2 are dimensionless empirical constants, and k_1 and k_2 are the drying constants. This model is best to describe the drying behavior of beet, onion, plum, pumpkin and stuffed pepper.

Two-term exponential model. The Two-term exponential model is a modification of the 2-term model, reducing the number of constants and modifying the indication of the constant form of the second exponential term. Erbay and Icier [12] emphasized that the constant "k_2" of the 2-term model is replaced by $(1 - a_1)$ at $t = 0$, then we have a moisture ratio $\overline{M}^* = 1$ This model can be expressed as follows:

$$\overline{M}^* = \frac{\overline{M} - M_e}{M_0 - M_e} = a_1 \exp(-K_1 t) + (1 - a_1)\exp(-K_2 a_1 t) \qquad (28)$$

Modified two-term model. The model involves a combination of Page and the 2-term model. The first part of the model is exactly like the Page model. However, describes more theoretically the model as a modified model of 2 terms with the inclusion of an empirical constant dimensionless "n". The model contains 5 constants and can be referred as a complex model.

$$\overline{M}^* = \frac{\overline{M} - M_e}{M_0 - M_e} = a_1 \exp(-K_1 t^n) + a_2 \exp(-K_2 t^n) \qquad (29)$$

Modified Midilli's model. This model is a modification of the Midilli's model. This mathematical model is expressed as follows:

$$\overline{M}^* = \frac{\overline{M} - M_e}{M_0 - M_e} = a_1 \exp(-K_1 t) + a_2 \qquad (30)$$

2.2.2 Empirical Models

Empirical models provide a direct relation between the average content moisture and drying time. The main limitation to the application of empirical models in thin-layer drying is that they do not follow the theoretical fundamentals of the drying processes in the form of a kinetic relation between the velocity constant and moisture concentration, thus giving imprecise parameter values. The following models are best suited to describe adequately the drying kinetics of some materials:

Aghbashlo and others model. Aghbashlo et al. [53] proposed a model that effectively describes the thin layer drying kinetics of biological materials. The model contains 2 dimensionless constants that depend on the absolute temperature of the drying system. However, there is no theoretical basis for this model:

$$\overline{M}^* = \frac{\overline{M} - M_e}{M_0 - M_e} = a_1 \exp\left(\frac{K_1 t}{1 + K_2}\right) \qquad (31)$$

where K_1 and K_2 are drying constants.

Wang and Singh model. This model was developed for the intermittent drying of wet biological material [54]. The model gives a good fit to the experimental data. However, this model has no physical or theoretical interpretation, hence its limitation.

$$\overline{M}^* = \frac{\overline{M} - M_e}{M_0 - M_e} = 1 + K_1 t + K_2 t^2 \qquad (32)$$

where K_1 and K_2 are model constants obtained from the experimental data.

Diamante and others model. Diamante et al. [55] proposed a new empirical model for the drying of biological materials. The equation of this model is given below:

$$\ln\left(-\ln\overline{M}^*\right) = a_1 + a_2(\ln t) + a_3(\ln t)^2 \tag{33}$$

where a_i are constants of the model. Again, this model lacks theoretical basis and physical interpretation.

Weibull's model. According to Tzempelikos et al. [49], this model was considered one of the most suitable empirical models and widely used in the literature. The model was, in fact, derived from experimental data, without physical or theoretical meanings. The equation related to this model is given by:

$$\overline{M}^* = \frac{\overline{M} - M_e}{M_0 - M_e} = a_1 - a_2\exp(-K_1 t^n) \tag{34}$$

where a_i are dimensionless model constants and k_1 is a drying constant.

Thompson's model. According to Pardeshi et al. [56], the Thompson's model is an empirical model obtained from experimental data, correlating the drying time as a function of the logarithm of the dimensionless moisture content. The model cannot describe the drying behavior of most materials because it has no theoretical basis and has no physical interpretation. The model can be expressed as follows:

$$t = a_1\ln\left(\overline{M}^*\right) + a_2\left[\ln\left(\overline{M}^*\right)\right]^2 \tag{35}$$

where a_i are dimensionless empirical constants.

Silva and others model. Silva et al. [57] proposed an empirical model for kinetic modeling of agricultural products. The equation of this model is given as follows:

$$\overline{M}^* = \frac{\overline{M} - M_e}{M_0 - M_e} = \exp\left(-a_1 t - a_2\sqrt{t}\right) \tag{36}$$

where a_1 and a_2 are fit parameters.

Peleg's model. This model has no physical meaning or theoretical interpretation. However, it was successfully applied in the production of the drying behavior of biological products. The mathematical equation related to this model is exposed as follows:

$$\overline{M}^* = \frac{\overline{M} - M_e}{M_0 - M_e} = 1 - \frac{t}{(a_1 + a_2 t)} \tag{37}$$

where a_1 and a_2 are dimensionless parameters of the model.

2.2.3 Theoretical Models

Non Phenomenological Models

The diffusion equation, for various geometries, has analytical solution for the average moisture content, whose overall form is given as follows:

$$\overline{M}^* = \frac{\overline{M} - M_e}{M_0 - M_e} = \sum_{n=1}^{\infty} A_n \exp(-B_n t) \tag{38}$$

where the values of A_n e B_n depend on the geometry of the body (plate, cylinder, sphere, parallelepiped, etc.), and boundary conditions (equilibrium, impermeable or convective).

In Eq. (38), the successive terms in each of the infinite convergent series decrease with increasing n and for long times, convergence can be obtained quickly. For sufficiently high values of t and equilibrium conditions at the surface of the solid, the first 5 terms dominate the series, and consequently the other terms in the series can be neglected. Further, for finite integer n, we have can write Eq. (38) as follows:

$$\overline{M}^* = \frac{\overline{M} - M_e}{M_0 - M_e} = \sum_{n=1}^{m} A_n \exp(-B_n t) \tag{39}$$

The value of m determines the accuracy of the $\overline{M}^*$ value calculated at each time instant. By analyzing Eq. (39) can we see, for example, that:

(a) If $m = 1$, $A_n = 1$ and $B_n = K_1$, this equation is reduced to Eq. (20);
(b) If $m = 1$, $A_n = a_1$, $B_n = K_1$ and $n = 1$, this equation is reduced to Eq. (21);
(c) Se $m = 1$, $A_n = 1$, and $n = 1$, this equation is reduced to Eq. (23); * Se $m = n$, $A_n = a_1$ and $B_n = K_1$, $i = 1, 2$ and 3, this equation is reduced to Eq. (24), and so on.

Then, most of the empirical semi-empirical models are derived from the diffusion model (non phenomenological theoretical model), and therefore their equations are approximations and variations of the diffusional model, depending on the number of terms used.

Herein, we start that the coefficients An and Bn depend on the shape of the body and the boundary conditions, which depend on the drying condition around the porous material. Therefore, when a specific equation is fitted to drying kinetics data of a particular product, the coefficients on this equation contain information of the external conditions (temperature, relative humidity, velocity, etc.). Therefore, it is perfectly acceptable and physical meaning that these coefficients can be considered constants or functions of the thermodynamic conditions and drying air velocity.

Phenomenological Models

Because of the limitations presented by the empirical and semi-theoretical/
semi-empirical methods, the literature has reported some theoretical models that
consider the phenomenology of the processes of mass loss and heating of a porous
material during the drying process based on the global capacitance method (lumped
analysis).

For the understanding of the global capacitance method consider a solid body
with arbitrary shape as shown in Fig. 2. The solid can receive (or to supply) a heat
and/or moisture flux per unit area on its surface and have internal generation of
mass and/or energy per unit volume uniformly distributed. Assuming that the
moisture and/or temperature of the solid is spatially uniform at any time during the
transient process, that is, the moisture and/or temperature gradients within the solid
are negligible, all mass and/or heat flux received (or supplied) and mass and/or heat
generated, will diffuse instantaneously through it.

Therefore, the global capacitance method admits a uniform distribution of mass
and or temperature within the solid at any instant, so that the temperature or
moisture content of the solid is given exclusively as a function of time [58].

In Fig. 2, $T\infty$ is the temperature of the external medium; h_c is the heat transfer
coefficient; h_m is the convective mass transfer coefficient; V is the volume of the
solid; S is the surface area of the solid; cp is the specific heat; M is the moisture
content of the product in any time interval; M_0 is the initial moisture content of the
product and Me is the equilibrium moisture content.

Applying a mass and energy balance in an infinitesimal element on the surface of
the solid, in any coordinate system, assuming constant thermo physical properties
and dimensional variations negligible, we obtain the following of mass and energy
conservation equations, respectively:

$$V\frac{d\overline{M}}{dt} = -M''S + \dot{M}V \qquad (40)$$

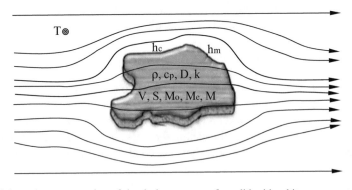

Fig. 2 Schematic representation of the drying process of a solid with arbitrary geometry

$$V\rho\frac{d\overline{\theta}}{dt} = -\frac{q''}{c_p}S + \frac{\dot{q}}{c_p}V \tag{41}$$

where ρ is the density of the solid; t is the time; M'' is the mass flux per unit area; $\dot{M}$ is the mass generation per unit volume; q'' is the heat flux per unit area; $\dot{q}$ is the heat generation per unit volume and $\overline{\theta}$ is the average temperature of the solid.

The quantities q'', M'', $\dot{q}$ and $\dot{M}$ may be positive or negative, and may also be constant or time dependent. Particularly with respect to energy, the quantity q'' can be, radiative, convective and evaporative or vapor heating. This formulation can be applied in regions of simultaneous heat and mass transfer. The particular case occurs when the two phenomena are completely independent (uncoupled phenomena). The two phenomena are coupled when absorption and desorption in a specific region are accompanied by thermal effects.

From the information given herein, and knowing that in many physical situations and operational conditions there is moisture and temperature gradients inside the solid, in what conditions can be applied the global capacitance method? The answer is related for a dimensionless parameter, well known as Biot number of transfer, which is defined by a relationship between the resistance to conduction inside the body and the resistance to convection at the surface thereof, as follows:

$$Bi = \frac{R_{Cond}}{R_{Conv}} = \frac{\frac{L_1}{\Gamma^\phi S}}{\frac{1}{hS}} = \frac{hL_1}{\Gamma^\phi} \tag{42}$$

where Γ^ϕ can be thermal conductivity or mass diffusion coefficient, and L1 is a characteristic length of the body, for example, the mathematical relation volume per surface area of the solid (V/S).

The Biot number play important role in the diffusion problems involving convective effects in the borders. For situation in that $Bi \ll 1$, the experimental results suggest a reasonable uniform distribution of moisture or temperature inside the body at any time t, in the transient process. It can be noticed that for the analysis of mass and thermal diffusion problem, one must calculate the number of Biot and, once this is less than 0.1, the error associated with the use of the method of global capacitance is small. Obviously, this statement is dependent upon of the geometry of the solid and like this parameter is defined. For example, the lumped parameters models are applied to Biot number of mass transfer smaller than 10 and Biot number of heat transfer smaller than 1.5 [59].

For physical situation where heat and mass transfer occur, the heat and mass fluxes per unit area as given by the following equations, respectively:

$$q'' = h_c(\theta - \theta_\infty) \tag{43}$$

$$M'' = h_m(M - M_e) \tag{44}$$

where h_m and h_c are the convective heat transfer coefficient and convective mass transfer coefficient, respectively. The determination of heat and mass transfer coefficients can be realized in two ways. The first method is based on finding the appropriate analytical relations from empirical data or by approximate solution of differential equations that describe the heat and mass transfer. The second method is based on the theory of similarity. The description of this theory can be found in books on heat and mass transfer and fluid mechanics.

The heat and mass transfer between the wet material and the drying agent depends on many external parameters whose influence is included in appropriate dimensionless numbers. The general form of this type of equations for heat transfer is as follows:

(a) Forced convection

$$Nu = f_1(Re, Pr, Gu) \tag{45}$$

(b) Free convection

$$Nu = f_2(Gr, Pr) \tag{46}$$

Similarly, for mass transfer, we can write:

(a) Forced convection

$$Sh = f_3(Re, Sc, Gu) \tag{47}$$

(b) Free convection

$$Sh = f4 \, (Gr', Sc) \tag{48}$$

where in these equations Nu is the Nusselt number, Sh is the Sherwood number, Re represent the Reynolds number, Sc is the Schmidt number, Pr is the Prandtl number, Gr is the Grashof number for heat transfer, Gr' is the Grashof number for mass transfer and Gu is Gukhman's number [10].

The following will be described some concentrated phenomenological models reported in the literature applied to homogeneous and heterogeneous solids with arbitrary shape.

(a) Homogeneous solids

To describe the heat and mass transfer in homogeneous solids (not hollow) (Fig. 2), Lima et al. [60], Silva [61] and Lima et al. [62] present a mathematical modelling based on the conservation laws of energy and mass, as follows:

- **Analysis of mass transfer**

$$V \frac{d\overline{M}}{dt} = -h_m S \left(\overline{M} - \overline{M}_e \right) + \dot{M} V \tag{49}$$

- **Analysis of heat transfer**

$$\rho V c_p \frac{d\overline{\theta}}{dt} = h_c S \left(\overline{\theta}_\infty - \overline{\theta} \right) + \rho_s V \frac{d\overline{M}}{dt} \left[h_{fg} + c_v \left(\overline{\theta}_\infty - \overline{\theta} \right) \right] + \dot{q} V \tag{50}$$

More recently, Silva et al. [63] and Lima [64] have applied the mathematical modeling proposed by Silva [61] in the drying of hollow homogeneous solids (ceramic tubes), as established in Fig. 3. The basic difference is related to surface area of heat and mass transfer, which include the external and internal surface area of the solid.

(b) Heterogeneous solids

To describe the heat and mass transfer in heterogeneous solids (not hollow) and with arbitrary shape, Almeida [65] presents a mathematical modelling based on the conservation laws of energy and mass, as follows (Fig. 4).

- **Analysis of mass transfer**

For the solid 1 we can write, the following mass balance:

$$\frac{D_1 S_1}{\Delta X_1} \left(\overline{M}_2 - \overline{M}_1 \right) = V_1 \frac{d\overline{M}_1}{dt} \tag{51}$$

For the solid 2, we can have that:

$$h_m S_2 \left(\overline{M}_e - \overline{M}_2 \right) + \frac{\rho_{1D_1 S_1}}{\rho_2 \Delta X_1} \left(\overline{M}_1 - \overline{M}_2 \right) = V_2 \frac{d\overline{M}_2}{dt} \tag{52}$$

- **Analysis of heat transfer**

For the solid 1, we can write, the following equation:

Fig. 3 Representative
scheme of the drying process
of a hollow homogeneous
solid with arbitrary geometry

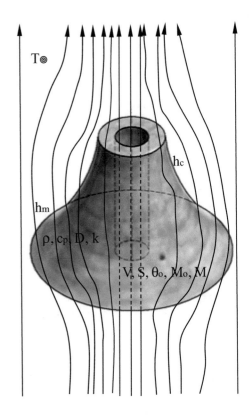

Fig. 4 Schematic
representation of a
heterogeneous solid
composed of two different
materials

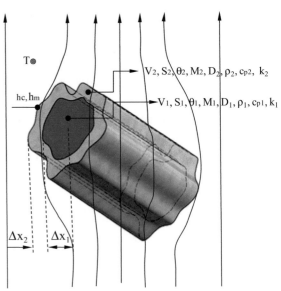

$$\frac{K_1 S_1}{\Delta X_1} (\overline{\theta}_2 - \overline{\theta}_1) = \rho_1 V_1 C_{P1} \frac{d\overline{\theta}_1}{dt} \tag{53}$$

For the solid 2, the energy conservation equation will be given as follows:

$$h_c S_2 (T_\infty - \overline{\theta}_2) + \frac{K_1 S_1}{\Delta X_1} (\overline{\theta}_1 - \overline{\theta}_2) = \rho_2 V_2 C_{P2} \frac{d\overline{\theta}_2}{dt} \tag{54}$$

where, in these equations, V is the volume, S is the surface area, C_p is the specific heat, K is the thermal conductivity, $\overline{M}$ is the average moisture content, $\overline{\theta}$ is the average temperature, and h_m and h_c are, respectively, the convective mass and heat transfer coefficients, D is the mass diffusion coefficient, and ρ is the density of the material.

3 Advanced Modeling to Describe the Drying of Hollow Homogeneous Solids

3.1 Basic Information

Figure 5 illustrates a hollow solid of arbitrary shape (wet and cold) with a fluid flowing around it (hot and dry). In this figure ρ is the density of the homogeneous solid; T_∞ is the temperature of the external medium; h_{c1} and h_{c2} are the of internal and external convective heat transfer coefficients, respectively; h_{m1} and h_{m2} are the internal and external convective mass transfer coefficients, respectively; V is the volume of the homogeneous solid; S_1 and S_2 are the internal and external surface area of the homogeneous solid, respectively; c_p is the specific heat, and $\overline{M}$ is the average moisture content on dry basis, and $\overline{\theta}$ is the average temperature of the product in any time instant.

Applying a mass and energy balance in an infinitesimal element at the surface of the solid, in any coordinate system, assuming constant thermo-physical properties and negligible dimensional variations, we have the following mass and energy conservation equations, respectively.

$$\rho_s V \frac{d\overline{M}}{dt} = -M'' S + \dot{M} \rho_s V \tag{55}$$

$$\rho_u V c_p \frac{d\overline{\theta}}{dt} = -q'' S + \dot{q} V \tag{56}$$

where the subscript s and u represent dry solid and wet solid, respectively, t is the time; M'' is the water mass flux per unit area; $\dot{M}$ is the moisture generation; q'' is the heat flux per unit area, and $\dot{q}$ is the heat generation per unit volume.

Fig. 5 Schematic
representative of the drying
process of a hollow
homogeneous solid with
arbitrary geometry

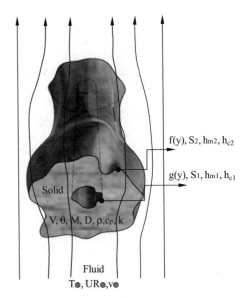

f(y), S₂, h$_{m2}$, h$_{c2}$

g(y), S₁, h$_{m1}$, h$_{c1}$

Solid

V, θ, M, D, ρ,c$_p$, k

Fluid
T⊕, UR⊕,v⊕

3.2 Mathematical Modeling of Heat and Mass Transfer

To model the drying process of solids with arbitrary shape (Fig. 4), the following
considerations were adopted:

(a) The solid is homogeneous and isotropic;
(b) The distribution of moisture and temperature inside the solid are uniform;
(c) The thermo-physical properties are constant throughout the process;
(d) The drying phenomenon occurs by conduction of heat and mass inside the solid
    and by convection of heat and mass and evaporation at the surface thereof.
(e) The solid have constant dimensions throughout the process.

3.2.1 Analysis of Mass Transfer

For mass transfer, M'' can be treated in the form of mass convection while, $\dot{M}$ can
be given, for example, by mass generation due to chemical reactions. Assuming that
the drying process occurs by convection M'' and $\dot{M}$ is constant, we can have by
replacement, into Eq. (55) the following equation for mass transfer:

$$V\frac{d\overline{M}}{dt} = -h_{m1}S_1\left(\overline{M} - \overline{M}_e\right) - h_{m2}S_2\left(\overline{M} - \overline{M}_e\right) + \dot{M}\,V \qquad (57)$$

where M_e e is the equilibrium moisture content on a dry basis

Using the initial condition $M(t = 0) = M_0$, separating the variables of Eq. (57) and integrating it since the initial condition, we have as a result, the following equation for determination of the average moisture content of the solid along the drying process:

$$\frac{(\overline{M} - \overline{M}_e) - \dfrac{\dot{M} V}{h_{m1} S_1 + h_{m2} S_2}}{(\overline{M}_0 - \overline{M}_e) - \dfrac{\dot{M} V}{h_{m1} S_1 + h_{m2} S_2}} = \mathrm{Exp}\left[\left(\frac{-h_{m1} S_1 - h_{m2} S_2}{V}\right) t\right] \tag{58}$$

3.2.2 Analysis of Simultaneous Heat and Mass Transfer

For the analysis of heat transfer, one can make analogy to mass transfer and to consider that at the surface of the solid, simultaneously occurs the phenomena of thermal convection, evaporation and heating of the vapor produced. Therefore, Eq. (56), can be written as follows:

$$\rho_u V c_p \frac{d\overline{\theta}}{dt} = h_{c1} S_1 (\overline{\theta}_\infty - \overline{\theta}) + h_{c2} S_2 (\overline{\theta}_\infty - \overline{\theta}) + \rho_s V \frac{d\overline{M}}{dt} \left[h_{fg} + c_v (\overline{\theta}_\infty - \overline{\theta})\right] + \dot{q} V \tag{59}$$

where, c_v is the specific heat of the vapor; h_{fg} is the latent heat of water vaporization; $\overline{\theta}_\infty$ is the temperature of the external medium; $\overline{\theta}_0$ is the initial temperature of the solid; $\overline{\theta}$ is the instantaneous average temperature of the solid; ρ_s is the specific mass of the dry solid; h_c is the convective heat transfer coefficient.

Equation (59), is an ordinary differential equation of first order, non-linear and non-homogeneous, and therefore cannot be resolved in analytical form. Thus, for simplification of Eq. (59), we assume negligible the energy required to heat the water vapor, from the temperature at the surface of the solid until the temperature of the fluid. So, after this simplification and performing the substitution of Eqs. (57) and (58) into Eq. (59), we have as a result:

$$\rho V c_p \frac{d\overline{\theta}}{dt} = (h_{c1} S_1 + h_{c2} S_2)(\overline{\theta}_\infty - \overline{\theta})$$
$$+ \rho_s h_{fg} \left\{ (-h_{m1} S_1 - h_{m2} S_2) \left[\left[(\overline{M}_0 - \overline{M}_e) - \frac{\dot{M} V}{h_{m1} S_1 + h_{m2} S_2} \right] \right.\right.$$
$$\left. \mathrm{Exp}\left[\left(\frac{-h_{m1} S_1 - h_{m2} S_2}{V}\right) t\right] + \frac{\dot{M} V}{h_{m1} S_1 + h_{m2} S_2} \right] + \dot{M} V \Bigg\} + \dot{q} V \tag{60}$$

or yet

$$\frac{d\bar{\theta}}{dt} = \left(\frac{(h_{c1}S_1 + h_{c2}S_2)}{\rho V c_p}\right)(\bar{\theta}_\infty - \bar{\theta})$$
$$+ \frac{\rho_s h_{fg}}{\rho V c_p}\left\{\left[(-h_{m1}S_1 - h_{m2}S_2)(\overline{M}_0 - \overline{M}_e) + \dot{M}V\right]\right.$$
$$\left.\text{Exp}\left[\left(\frac{-h_{m1}S_1 - h_{m2}S_2}{V}\right)t\right]\right\} + \frac{\dot{q}}{\rho c_p} \tag{61}$$

Admitting $y = \bar{\theta}_\infty - \bar{\theta}$, then $\frac{dy}{dt} = -\frac{d\bar{\theta}}{dt}$. Therefore, Eq. (61) can be written as follows:

$$y' + a = -be^{-ct} - d \tag{62}$$

where

$$a = \frac{(h_{c1}S_1 + h_{c2}S_2)}{\rho V c_p} \tag{63}$$

$$b = \frac{\rho_s h_{fg}}{\rho c_p}\left[(-h_{m1}S_1 - h_{m2}S_2)(\overline{M}_0 - \overline{M}_e) + \dot{M}V\right] \tag{64}$$

$$c = \frac{-h_{m1}S_1 - h_{m2}S_2}{V} \tag{65}$$

$$d = \frac{\dot{q}}{\rho c_p} \tag{66}$$

Using the initial condition $\bar{\theta}(t = 0) = \bar{\theta}_0$, and solving Eq. (62) we obtain the following equation for determination of the average temperature of the solid along the drying process:

$$\bar{\theta} = \bar{\theta}_\infty - \left[(\bar{\theta}_\infty - \bar{\theta}_0) + \left(\frac{b}{a - c} + \frac{d}{a}\right)\right]e^{-at} + \left(\frac{b}{a - c}e^{-ct} + \frac{d}{a}\right) \tag{67}$$

where the parameters a, b, c and d are given by Eqs. (63)–(66), respectively.

3.2.3 Volume and Surface Area of the Solid with Arbitrary Shape

To find the volume of a solid with arbitrary shape was used the method of circular rings applied to solids of revolution [66]. This method consists of assuming that f and g (Fig. 6) are non-negative functions on the interval [y_1, y_2], such that $f(y) \geq g(y)$ for all values of y in the interval [y_1, y_2], and let R to be the plane

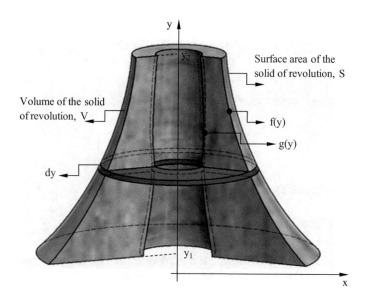

Fig. 6 a Plane region and **b** solid of revolution

region bounded by the graphs of f and g between $y = y_1$ and $y = y_2$. Let S to be the solid generated by the revolution of R around the x-axis (Fig. 6b, c).

Following, consider an infinitesimal volume dV of the solid as pictured in the shaded volume consisting of a circular ring of infinitesimal thickness dy, which is perpendicular to the axis of revolution and centered in a point of the y axis. The base of this circular ring is the region between the two concentric circles of radius $f(y)$ and $g(y)$, so that area of this base is $\pi f(y)^2 - \pi g(y)^2$ square units. Then, the volume of the solid will be given as follows:

$$V = \int_{y1}^{y2} \pi[f(y)^2 - g(y)^2]dy \qquad (68)$$

The surface area of the solid of revolution studied in this research was obtained by the revolution generated by the rotation of the portion of the graph of the continuous and non-negative functions $f(y)$ and $g(y)$ between the lines $y = y_1$ and $y = y_2$ around the y axis [66]. Then, the internal and external surface areas of the hollow solid are given as follows, respectively:

$$S_1 = \int_{y1}^{y2} 2\pi g(y)\sqrt{1 + [g'(y)]^2}dy \qquad (69)$$

$$S_2 = \int\limits_{y1}^{y2} 2\pi f(y)\sqrt{1 + [f'(y)]^2}\,dy + \pi\left[f(y_2)^2 - g(y_2)^2\right] + \pi\left[f(y_1)^2 - g(y_1)^2\right] \quad (70)$$

Thus, the total surface area of the hollow solid will be given by:

$$S = S_1 + S_2 \quad (71)$$

4 Application: Drying of Hollow Solids

4.1 The Geometry of the Solid

According to Fig. 6, in this research it was adopted the following functions f and g, in order to define the shape of the solids to be studied:

$$f(y) = \left\{ a^m\left[1 - \left(\frac{y}{b}\right)^2\right] \right\}^{\frac{1}{m}} \quad (72)$$

which corresponds to the contour of an ellipse, where $\underline{a}$ and $\underline{b}$ are the major and minor semi-axes of the ellipse, respectively, and

$$g(y) = a' = \text{constant} \quad (73)$$

Further, in Fig. 6, $y_1 = 0$, $y_2 = b'$ and m are constants that define the shape of the body.

4.2 Process Parameters and Cases in Study

Table 2 summarizes the thermo-physical properties of the solid and fluid, and Table 3 contain information of the geometric parameters of the solid, which were used in the simulations.

4.3 Heat and Mass Transfer Analysis

Figures 7, 8, 9 and 10 illustrate the geometries considered in this study. The different forms were obtained by varying the parameters a', b' and m, as described in Table 3.

Table 2 Parameter of the materials used in the simulations

Parameter	Material	
	Solid	Air
Wet solid density (kg/m^3)	640	–
Dry solid density (kg/m^3)	550	–
Specific heat of wet solid (J/kg K)	1600	–
Latent heat of water vaporization in the solid (J/kg)	2.333×10^3	–
Mass diffusion coefficient (m^2/s)	1×10^{-9}	–
Thermal conductivity (W/m^2 K)	0.833	–
Initial temperature (°C)	25	70
Final temperature (°C)		70
Initial moisture content $\left(\text{kg}_{\text{water}} / \text{kg}_{\text{dry solid}} \right)$	0.20	–
Final moisture content $\left(\text{kg}_{\text{water}} / \text{kg}_{\text{dry solid}} \right)$	0.01	–
Internal convective mass transfer coefficient (m/s)	–	1×10^{-9}
External convective mass transfer coefficient (m/s)	–	5×10^{-9}
Internal convective heat transfer coefficient (W/m^2 K)	–	4
External convective heat transfer coefficient (W/m^2 K)	–	8

Table 3 Values of the geometries of the solid of revolution

Case	a(m)	b(m)	m	b'(m)	a'(m)
1	0.05	0.20	0.50	0.025	0.005
2	0.05	0.20	0.50	0.050	0.005
3	0.05	0.20	0.50	0.075	0.005
4	0.05	0.20	0.50	0.100	0.005

In Table 4, the following geometric parameters of the solid in study: area, volume, and area/volume ratio, can be observed.

In this work, a comparison was made between the drying kinetics of four solids with different types of geometries, becoming possible a better understand of the heat and mass transfer during the drying process of hollow solids.

Figure 11 shows the drying kinetics for the cited cases. After analysis of these figures, we can see that the moisture content at the beginning of the drying process is exponentially reduced until reaching equilibrium moisture content (hygroscopic equilibrium) at the end of the process. This behavior demonstrates that the drying process happens only in the falling drying rate period (non-existence of the constant drying rate period). The stage of falling drying rate is governed by internal migration of moisture, being characterized by a decline in the drying rate. Since the moisture of the product decreases during drying, the rate of internal moisture movement also decreases and thus the drying rate drops rapidly.

Figures 11 illustrates the influence of parameter b', which represents the height variation of the solid. After analysis of this figures is verified that increasing the

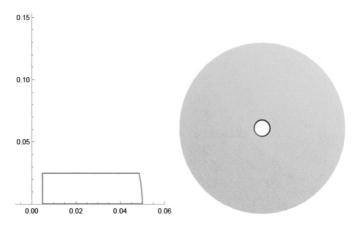

Fig. 7 Geometry of the solid for the case 1 (a = 0.05; b = 0.2; m = 0.5; b' = 0.025 and a' = 0.005)

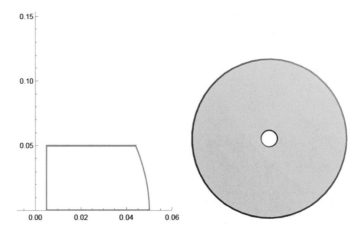

Fig. 8 Geometry of the solid for the case 2 (a = 0.05; b = 0.2; m = 0.5; b' = 0.050 and a' = 0.005)

value of parameter b' (height of the revolution solid), the solid dries more slowly. This occurs independently of the curvature of the solid (parameter m).

We notice that the influence of the parameter b', on the drying kinetics of the hollow solids is more significant than the parameter a' (hole diameter), fixed the shape of the body (m constant). However, the effect of the parameter a' is less influential when the parameters m and b' assume smallest values.

This can be explained because the shape of the solid directly influences the drying rate, that is, the area/volume ratio is a predominant variable in the evaluation process. The higher the area/volume ratio the faster the solid loses moisture.

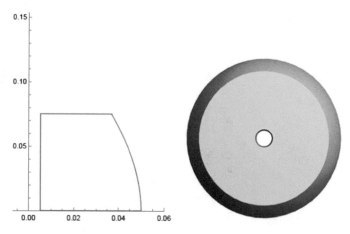

Fig. 9 Geometry of the solid for the case 3 (a = 0.05; b = 0.2; m = 0.5; b' = 0.075 and a' = 0.005)

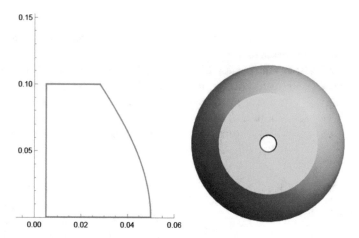

Fig. 10 Geometry of the solid for the case 4 (a = 0.05; b = 0.2; m = 0.5; b' = 0.100 and a' = 0.005)

Table 4 Values of area, volume, and area/volume ratio for the cases in study

Case	S_1 (m^2)	S_2 (m^2)	S (m^2)	V (m^3)	S\V (m^{-1})
1	0.00078540	0.0228636	0.0236490	0.000190353	124.238
2	0.00157080	0.0289692	0.0305400	0.000357834	85.3468
3	0.00235619	0.0338127	0.0361689	0.000485779	74.4554
4	0.00314159	0.0374474	0.0405890	0.000567978	71.4622

Fig. 11 Moisture content as
a function of time for different
solids of revolution

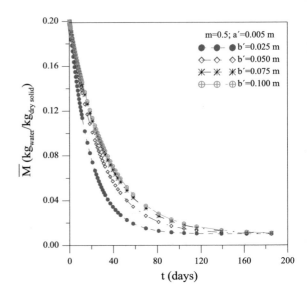

The effect of the parameter m in the drying process is small when compared to
the other geometric parameters.

Figure 12 illustrates the heating kinetics, which corresponds to the results of the
temperature behavior as a function of time, for the all geometries studied.

Analyzing these figures, it is noticed that the solids reach the temperature of the
drying air (thermal equilibrium) faster compared to mass transfer. This behavior is
due to the fact that the mass diffusion coefficient at the solid is much smaller than

Fig. 12 Temperature as a
function of time for different
solids of revolution

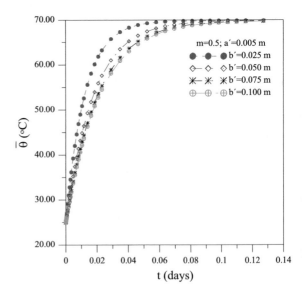

the thermal diffusivity. When the solid enters in thermal equilibrium with the drying air, the mass transfer process inside the solid occurs in the isothermal condition. This occurs from the first hour of the process.

The area/volume ratio of the material is a factor of great influence in the process of heating, since the higher this ratio the faster the process will happen.

When the surface area is large compared to the volume, the heat and mass transfer area will be increased, whereas the distance traveled by the heat and mass flux inside the material until the solid surface will decrease, thus we have an increase in heating and mass transfer rates.

Despite the problems presented due to the rapid drying, and the necessity of a rigorous control of the process, it was observed that the drying time in this research was very long. The global capacitance method involves a strict control with respect to the problems related to the drying of wet materials, but it has a great disadvantage, the cost of the process would be high when compared to the drying methods currently used.

About the proposed model, the following general comment can be summarized:

(a) The proposed mathematical model is versatile and can be used in different operating conditions, with constant, impermeable or convective boundary conditions, and also with other boundary conditions, under small modifications.
(b) Reasonable degree of approximation, allowing obtain good adjustments with experimental data, and parameters estimation with high accuracy.
(c) It can be used in different geometries and types of materials without nature restrictions (fruits, vegetables, grains, clay products, wood, etc.);
(d) It can be used in different geometries and types of materials without nature restrictions (fruits, vegetables, grains, clay products, wood, etc.);
(e) Ease of extension to other physical situations involving heat and mass transport, such as in the use of infrared or microwave drying;
(f) High stability and low numerical computational time when compared with other solution methods of differential equations.

5 Concluding Remarks

In this chapter, heat and mass transfer in wet solids has been explored, with particular reference to drying of irregularly-shape porous solids. Interest in this physical problem is motivated by the great importance in many practical situations related to moisture removal by heating.

A consistent and advanced lumped phenomenological model (thin layer drying model) applied to solid with complex shape is proposed, and all mathematical formalism for obtain the exact solution of the governing equation is presented. Application has been done to hollow solids.

From the simulations and results presented, it is possible to conclude that:

(a) The proposed mathematical modeling used was adequate, and it can be applied to different process such as: drying, wetting, cooling and heating of solids with arbitrary shape.
(b) The heat transfer occurs more quickly than the mass transfer, which contributes to the solids to reach the equilibrium temperature in a shorter time.
(c) The surface area of the product that is exposed to drying and its volume plays important role in drying process. The higher the area/volume relationships the faster the drying.

The drying process at higher velocity may interfere in the quality of the solid with respect to defects such as cracks, fissures, color, etc. However, if the drying velocity is slower, the process becomes undesirable, because the excessive expenditure of energy and low productivity. Thus, it is necessary to find best condition that provides the better relation between cost and benefit.

Acknowledgements The authors thank to CNPq, FINEP and CAPES (Brazilian Research Agencies) for financial support and to the authors referred in this text that contributed for improvement of this work.

References

1. Park, K.J., Brod, F.P.R.: Comparative study of grated coconut (*Cocos nucifera*) drying using vertical and horizontal dryers. In: Inter-American Drying Conference (IADC), Itu, Brazil, **B**, pp. 469–475 (1997)
2. Steffe, J.F., Singh, R.P.: Liquid diffusivity of rough rice components. Trans. ASAE **23**(3), 767–774 (1980)
3. Fortes, M., Okos, M.R.: Drying theories: their bases and limitations as applied to foods and grains. In: Advances in Drying, vol. 1, pp. 119–154. Hemisphere Publishing Corporation, Washington (1980)
4. Alvarenga, L.C., Fortes, M., Pinheiro Filho, J.B., Hara, T.: Moisture transport inside the black bean grains under drying conditions. Rev. Bras. de Armazenamento **5**(1), 5–18 (1980) (in Portuguese)
5. Mariz, T.F.: Drying of cotton seed shell in fixed bed. Master dissertation in Chemical Engineering, Federal University of Paraiba, Campina Grande, Brazil (1986) (in Portuguese)
6. Keey, R.B.: Drying of Loose and Particulate Materials. Hemisphere Publishing Corporation, New York (1992)
7. Lima, A.G.B.: Drying study and design of silkworm cocoon dryer. Master dissertation in Mechanical Engineering, Federal University of Paraiba, Campina Grande, Brazil (1995) (in Portuguese)
8. Ibrahim, M.H., Daud, W.R.W., Talib, M.Z.M.: Drying characteristics of oil palm kernels. Drying Technol. **15**(3–4), 1103–1117 (1997)
9. Lima, A.G.B.: Diffusion phenomena in prolate spheroidal solids: case studies: drying of bananas. Doctorate thesis in Mechanical Engineering, State University of Campinas, Campinas, Brazil (1999) (in Portuguese)
10. Strumillo, C., Kudra, T.: Drying: Principles, Science and Design. Gordon and Breach Science Publishers, New York (1986)
11. Brooker, D.B., Bakker-Arkema, F.W., Hall, C.W.: Drying and Storage of Grains and Oilseeds. AVI Book, New York (1992)

12. Erbay, Z., Icier, F.: A review of thin-layer drying of foods: theory, modeling, and experimental results. Crit. Rev. Food Sci. Nutr. **50**(5), 441–464 (2010)
13. Luikov, A.V.: Heat and Mass Transfer in Capillary Porous Bodies. Pergamon Press, New York (1966)
14. Sarker, N.N., Kunze, O.R., Stroubolis, T.: Finite element simulation of rough rice drying. Drying Technol. **12**(4), 761–775 (1994)
15. Zogzas, N.P., Maroulis, Z.B.: Effective moisture diffusivity estimation from drying data: a comparison between various methods of analysis. Drying Technol. **14**(7–8), 1543–1573 (1996)
16. Liu, J.Y., Simpson, W.T.: Solutions of diffusion equation with constant diffusion and surface emission coefficients. In: Inter-American Drying Conference (IADC), Itu, Brazil, **A**, pp. 73–80 (1997)
17. Freire, E.S., Chau, K.V.: Simulation of the drying process of fermented cacao beans. In: Inter-American Drying Conference (IADC), Itu, Brazil, **B**, pp. 356–363 (1997)
18. Baroni, A.F., Hubinger, M.D.: Drying of onion: effects of pre-treatment on moisture transport. In: Inter-American Drying Conference (IADC), Itu, Brazil, **B**, pp. 419–426 (1997)
19. Sabadini, E., Carvalho Jr., B.C., Sobral, P.J.A., Hubinger, M.D.: Mass transfer and diffusion coefficient determination in salted and dried meat pieces. In: Inter American Drying Conference (IADC), Itu, Brazil, **B**, pp. 441–447 (1997)
20. Quintana-Hernandez, P., Rodrigues-Ramirez, J., Mendes-Lagunas, L., Cornejo-Serrano, L.: Humidity diffusion within sugarcane fibers. In: Inter-American Drying Conference (IADC), Itu, Brazil, **B**, pp. 538–542 (1997)
21. Oliveira, V.A.B., Lima, A.G.B.: Unsteady state mass diffusion prolate spheroidal solids: an analytical solution. In: Inter-American Drying Conference (IADC), Boca del Rio, Mexico, pp. 163–172 (2001)
22. Carmo, J.E.F., Lima, A.G.B.: Modelling and simulation of mass transfer inside the oblate spheroidal solids. In: Inter-American Drying Conference (IADC), Boca del Rio, Mexico, pp. 173–183 (2001)
23. Nascimento, J.J.S., Belo, F.A., Lima, A.G.B.: Simultaneous moisture transport and shrinkage during drying of parallelepiped solids. In: Inter-American Drying Conference (IADC), Boca del Rio, Mexico, pp. 535–544 (2001)
24. Karathanos, V.T., Vagenas, G.K., Saravacos, G.D.: Water diffusivity in starches at high temperatures and pressures. Biotechnol. Prog. **7**(2), 178–184 (1991)
25. Fortes, M., Okos, M.R.: A non-equilibrium thermodynamics approach to transport phenomena in capillary porous media. Trans. ASAE **24**, 756–760 (1981)
26. Ozdemir, M., Devres, Y.O.: The thin-layer drying characteristics of hazelnuts during roasting. J. Food Eng. **42**(4), 225–233 (2000)
27. Panchariya, P.C., Popovic, D., Sharma, A.L.: Thin-layer modeling of black tea drying process. J. Food Eng. **52**(4), 349–357 (2002)
28. Akpinar, E.K.: Determination of suitable thin-layer drying curve model for some vegetables and fruits. J. Food Eng. **73**(1), 75–84 (2006)
29. Doymaz, I.: The kinetics of forced convective air-drying of pumpkin slices. J. Food Eng. **79**(1), 243–248 (2007)
30. Raquel, P.F., Susana, P., Maria, J.B.: Study of the convective drying of pumpkin (*Cucurbita maxima*). Food Bioprod. Process. **89**(4), 422–428 (2011)
31. Henderson, S.M.: Progress in developing the thin-layer drying equation. Trans. ASAE **17**(6), 1167–1172 (1974)
32. Bruce, D.M.: Exposed layer barley drying, three models fitted to new data up to 150°C. J. Agric. Eng. Res. **32**(4), 337–348 (1985)
33. Santos, G.M.: Study of thermal behavior of a tunnel kiln applied to red ceramic industry. Master dissertation in Mechanical Engineering, Federal University of Santa Catarina, Florianópolis, Brazil (2001) (in Portuguese)
34. Nishikawa, T., Gao, T., Hibi, M., Takatsu, M., Ogawa, M.: Heat transmission during thermal shock testing of ceramics. J. Mater. Sci. **29**(1), 213–219 (1994)

35. Janjai, S., Lamlert, N., Mahayothee, B., Bala, B.K., Precoppe, M., Muller, J.: Thin-layer drying of peeled longan (*Dimocarpus longan* Lour.). Food Sci. Technol. Res. **17**(4), 279–288 (2011)
36. Akpinar, E.K.: Mathematical modeling of thin-layer drying process under open sun of some aromatic plants. J. Food Eng. **77**(4), 864–870 (2006)
37. Babalis, S.J., Papanicolaou, E., Kyriakis, N., Belessiotis, V.G.: Evaluation of thin-layer drying models for describing drying kinetics of figs (*Ficus carica*). J. Food Eng. **75**(2), 205–214 (2006)
38. Menges, H.O., Ertekin, C.: Mathematical modeling of thin-layer drying of golden apples. J. Food Eng. **77**(1), 119–125 (2006)
39. Vega, A., Uribe, E., Lemus, R., Miranda, M.: Hot-air drying characteristics of aloe vera (*Aloe barbadensis*) and influence of temperature on kinetic parameters. LWT Food Sci. Technol. **40** (10), 1698–1707 (2007)
40. Saeed, I.E., Sopian, K., Abidin, Z.Z.: Drying characteristics of Roselle (1): mathematical modeling and drying experiments. Agric. Eng. Int. CIGR J. Manuscript FP 08 015. **X**, 1–25 (2008)
41. Fadhel, M.I., Abdo, R.A., Yousif, B.F., Zaharim, A., Sopian, K.: Thin-layer drying characteristics of banana slices in a force convection indirect solar drying. In: 6th IASME/ WSEAS International Conference on Energy and Environment: Recent Researches in Energy and Environment, Cambridge, England, pp. 310–315 (2011)
42. Kadam, D.M., Goyal, R.K., Gupta, M.K.: Mathematical modeling of convective thin-layer drying of basil leaves. J. Med. Plants Res. **5**(19), 4721–4730 (2011)
43. Rasouli, M., Seiiedlou, S., Ghasemzadeh, H.R., Nalbandi, H.: Convective drying of garlic (*Allium sativum* L.): part I: drying kinetics, mathematical modeling and change in color. Aust. J. Crop Sci. **5**(13), 1707–1714 (2011)
44. Akoy, E.O.: Experimental characterization and modeling of thin-layer drying of mango slices. Int. Food Res. J. **21**(5), 1911–1917 (2014)
45. Gan, P.L., Poh, P.E.: Investigation on the effect of shapes on the drying kinetics and sensory evaluation study of dried jackfruit. Int. J. Sci. Eng. **7**(2), 193–198 (2014)
46. Tzempelikos, D.A., Vouros, A.P., Bardakas, A.V., Filios, A.E., Margaris, D.P.: Case studies on the effect of the air drying conditions on the convective drying of quinces. Case Stud. Therm. Eng. **3**, 79–85 (2014)
47. Darıcı, S., Sen, S.: Experimental investigation of convective drying kinetics of kiwi under different conditions. Heat Mass Transf. **51**(8), 1167–1176 (2015)
48. Onwude, D.I., Hashim, N., Janius, R., Nawi, N., Abdan, K.: Computer simulation of convective hot air drying kinetics of pumpkin (*Cucurbita moschata*). In: 8th Asia-Pacific Drying Conference (ADC 2015) Kuala Lumpur, Malaysia, pp. 122–129 (2015)
49. Tzempelikos, D.A., Vouros, A.P., Bardakas, A.V., Filios, A.E., Margaris, D.P.: Experimental study on convective drying of quince slices and evaluation of thin-layer drying models. Eng. Agric. Environ. Food **8**(3), 169–177 (2015)
50. Kucuk, H., Midilli, A., Kilic, A., Dincer, I.: A review on thin-layer drying-curve equations. Drying Technol. **32**(7), 757–773 (2014)
51. Midilli, A., Kucuk, H., Yapar, Z.: A new model for single-layer drying. Drying Technol. **20** (7), 1503–1513 (2002)
52. Sacilik, K.: Effect of drying methods on thin-layer drying characteristics of hull-less seed pumpkin (*Cucurbita pepo* L.). J. Food Eng. **79**(1), 23–30 (2007)
53. Aghbashlo, M., Kianmehr, M.H., Khani, S., Ghasemi, M.: Mathematical modeling of thin-layer drying of carrot. Int. Agrophysics **23**(4), 313–317 (2009)
54. Wang, C.Y., Singh, R.P.A.: single layer drying equation for rough rice. ASAE American Society of Agricultural and Biological Engineers, St. Joseph, MI, Paper No 78-3001 (1978)
55. Diamante, L., Durand, M., Savage, G., Vanhanen, L.: Effect of temperature on the drying characteristics, colour and ascorbic acid content of green and gold kiwifruits. Int. Food Res. J. **17**(2), 441–451 (2010)

56. Pardeshi, I.L., Arora, S., Borker, P.A.: Thin-layer drying of green peas and selection of a suitable thin-layer drying model. Drying Technol. **27**(2), 288–295 (2009)
57. Silva, W.P., Silva, C.M.D.P.S., Gama, F.J.A.: Mathematical models to describe thin-layer drying and to determine drying rate of whole bananas. J. Saudi Soc. Agric. Sci. **13**(1), 67–74 (2014)
58. Incropera, F.P., De Witt, D.P.: Fundamentals of Heat and Mass Transfer. Wiley, New York (2002)
59. Parti, M.: Selection of mathematical models for drying grain in thin-layers. J. Agric. Eng. Res. **54**(4), 339–352 (1993)
60. Lima, A.G.B., Farias Neto, S.R., Silva, W.P.: Heat and mass transfer in porous materials with complex geometry: fundamentals and applications. In: Delgado, J.M.P.Q. (Org.) Heat and Mass Transfer in Porous Media. Series: Advanced Structured Materials, vol. 13, 1st edn, pp. 161–185. Springer, Heidelberg (2011)
61. Silva, J.B.: Drying of solids in thin-layer via lumped analysis: modeling and simulation. Master dissertation in Mechanical Engineering, Federal University of Campina Grande. Campina Grande, Brazil (2002) (in Portuguese)
62. Lima, L.A., Silva, J.B., Lima, A.G.B.: Heat and mass transfer during drying of solids with arbitrary shape: a lumped analysis. J. Braz. Assoc. Agric. Eng. (Engenharia Agrícola, Jaboticabal) **23**(1), 150–162 (2003) (in Portuguese)
63. Silva, V.S., Delgado, J.M.P.Q., Barbosa de Lima, W.M.P., Barbosa de Lima, A.G.: Heat and mass transfer in holed ceramic material using lumped model. Diffus. Found. **7**, 30–52 (2016)
64. Lima, W.M.P.B.: Heat and mass transfer in porous solids with complex shape via lumped analysis: modeling and simulation. Master dissertation in Mechanical Engineering, Federal University of Campina Grande. Campina Grande, Brazil (2017) (in Portuguese)
65. Almeida, G.S.: Heat and mass transfer in heterogeneous solids with arbitrary shape: a lumped analysis. Master dissertation in Mechanical Engineering, Federal University of Campina Grande, Brazil (2003) (in Portuguese)
66. Munem, M.A., Foulis, D.J.: Calculus, Guanabara Dois S.A., Rio de Janeiro, Brazil. **1** (1978) (in Portuguese)

Water Absorption Process in Polymer Composites: Theory Analysis and Applications

R. Q. C. Melo, W. R. G. Santos, Antonio Gilson Barbosa de Lima, W. M. P. B. Lima, J. V. Silva and R. P. Farias

Abstract Transport phenomena in porous media represent an important research area related to heat and mass transfer, and fluid flow fields. This chapter presents information about anomalous behaviour of moisture transient diffusion in vegetable fiber-reinforced composites materials. Composites reinforced with natural fibers are sensitive to influences from environmental agents such as moisture and temperature. Herein, topics related to theory, experiments, mathematical modeling and solution procedures, and technological applications are presented and discussed in detail. An advanced model that (Langmuir-type model) to describe water absorption in polymer composites and the analytical (Laplace transform technique) and numerical (finite-volume method) solutions of the governing equation has been obtained, considering constant thermo-physical properties. In the Langmuir model, moisture sorption can be explained by assuming that water exists in the free and bound phases inside the material. Application has been done to water uptake in

R. Q. C. Melo
Department of Materials Engineering, Federal University of Campina Grande,
Av. Aprígio Veloso, 882, Bodocongó, 58429-900 Campina Grande, PB, Brazil
e-mail: rafaelaquinto@live.com

W. R. G. Santos · A. G. Barbosa de Lima (✉) ·
W. M. P. B. Lima · J. V. Silva
Department of Mechanical Engineering, Federal University of Campina Grande,
Av. Aprígio Veloso, 882, Bodocongó, 58429-900 Campina Grande, PB, Brazil
e-mail: antonio.gilson@ufcg.edu.br

W. R. G. Santos
e-mail: wanessa.raphaella@yahoo.com.br

W. M. P. B. Lima
e-mail: wan_magno@gmail.com

J. V. Silva
e-mail: jvieira7@gmail.com

R. P. Farias
Department of Agriculture Science, State University of Paraiba,
Catolé do Rocha, PB 58884-000, Brazil
e-mail: dinasansil@yahoo.com.br

© Springer International Publishing AG 2018
J. M. P. Q. Delgado and A. G. Barbosa de Lima (eds.), *Transport Phenomena in Multiphase Systems*, Advanced Structured Materials 93,
https://doi.org/10.1007/978-3-319-91062-8_7

Caroá fiber-reinforced polymer composites. Results of the absorption kinetics and concentration distribution of water (free and trapped water molecules) within the material along the process are presented and analyzed. Predicted results compared to experimental data of average moisture content have shown that the model was effective for description of the phenomenon, allowing a better understanding about the effects of moisture migration mechanisms.

Keywords Composites · Vegetable fiber · Langmuir model · Analytical Numerical

1 Introduction

The materials, depending on their chemical and structural characteristics, can be divided in four categories: metals, ceramics, polymers and composites. Specifically, composite materials consist of the blending of two or more materials with distinct individual characteristics, which retain their identity in such a way that, they can be physically identified and exhibit an interface between them. The constituent that forms the composite is called the matrix, while the other constituent, called reinforcement, and determine the internal structure of the composite.

The matrix involves the reinforcement, protecting it against chemical and environmental attacks surrounding the composite, mechanical abrasion and transferring the load to the reinforcement, when the composite is mechanically requested. The reinforcement provides strength and stiffness to the composite, which may be fibrous or particulate. The fibers used as reinforcement are found in different forms and each configuration results in distinct properties, being they strongly dependent on the way the fibers are arranged inside the composites.

The region between the constituents of a composite is called the interface which, in turn, is a boundary delimiting the distinct phases between matrix and fiber, playing a very important role in determining the final properties of the composite. At the interface, there may be chemical and physical interactions between fiber and matrix, and there may also be voids and gases. The interface is the region where fiber and resin are chemically and mechanically combined. Depending on the processing conditions, chemical reactions, and volumetric changes causing stress, the resulting interface will be very complex. For good interfacial resistance, the surface of the fibers must be well wetted or surrounded by the resin [1–7]. The volumetric contraction phenomenon that the polymer matrix undergoes during its cure or solidification can be reduced by the addition of fibers [8].

Inadequate adhesion between the phases involved may lead to the initiation of interfacial faults, compromising the overall performance of the composite. In composites with polymer matrix, the fault should occur in the matrix. In practice, the adhesion is never perfect and the process of rupture is generated at the interface. Therefore, in most cases, the failure of the reinforced polymer occurs by shearing in the interfacial region. The failure occurs due to the weakness of the atomic or intermolecular bonds between the surface of the matrix and the surface of the reinforcement [9]. Due to the importance, several works have been reported in the literature on this topic: Jayamol et al. [10], Bledzki and Gassan [11], Joseph et al. [12, 13], Medeiros et al. [14], Nóbrega et al. [15], Agarwal et al. [16], Carvalho and Cavalcanti [17], Carvalho et al. [18], Uday et al. [19], Haneefa et al. [20], Costa et al. [21], Venkateshwaran et al. [22], Santos et al. [23].

Composite materials are increasingly being used instead of traditional materials, such as metals, ceramics and polymers, whose individual characteristics do not meet the growing demands for better performance, safety, economy and durability. Another remarkable feature of the composites is their versatility in the broad spectrum of physical, chemical and mechanical properties that can be obtained by combining different types of matrix and the various reinforcement options [24, 25].

Currently, synthetic fibers are widely used as reinforcement for polymer matrix composites. However, problems related to cost and the search for materials that do not offer environmental risks are challenges to be overcome. Thus, the study of other fibers of high resistance, low cost and less polluting becomes essential to meet the current technological needs. In general, vegetable fibers were used in the manufacture of ropes, yarns, carpets and other decorative products. However, due to the potential to replace petroleum products (or mineral fillers) and the growing ecological appeal in present day, vegetable fibers have returned to the scene with great intensity in the production of polymer composites. They present many advantages, such as, are from renewable and biodegradable sources, recyclable, light, strong, relatively durable, low cost and neutral in relation to CO_2 emissions [26, 27]. Based on the new environmental regulations and sustainability concepts, growing ecological, social and economic awareness, vegetable fibers have presented great potential for technological application [28, 29]. This will not only reduce waste disposal problems but will also reduce environmental pollution [30].

Polymeric composites reinforced with vegetal fibers, also called Green composites, are ecological and economically viable materials, which minimize the generation of pollution. They refer to the combination of fully degradable fibers, cellulosic materials and natural resins. In Brazil, the use of vegetable fibers as a polymeric reinforcement has not only a technological but also socioeconomic importance, since they are used by different communities for handicrafts and for the production of ropes. This contributes to regional development, enhancing the cultivation of fibers and ensuring greater demand for them, as well as avoiding rural exodus.

Plant fibers in comparison to glass fibers are very efficient in sound absorption, have low cost, are light, do not shatter in the event of accidents, are biodegradable and can be obtained using 80% less energy than the glass fibers [31, 32]. Despite of various advantages presents herein, the following disadvantages and limitations can be mentioned: high moisture absorption, low thermal stability, lower mechanical properties than synthetic fibers, marked variability in mechanical properties, low dimensional stability, high sensitivity to environmental effects such as variations in temperature and humidity, soil, harvest time, post-harvest processing, relative location in the plant body and processing temperatures [33]. The maximum application temperature of the vegetable fibers is relatively low, around 200 °C, which causes the mechanical properties of the fibers to be below their properties at room temperature. For composites with thermosetting resins, this characteristic is not limiting, since curing of the resins usually takes place at temperatures below 200 °C [34].

2 Water Sorption by Polymer Composites

2.1 Fundamentals

Because the hydrophilic nature of plant fibers, the water-absorbing capacity of the vegetable fiber reinforced composites is higher than the synthetic fiber-reinforced composites. The hydrophilicity and high capillarity characteristics of natural fibers, their use in predominantly hydrophobic matrix polymer composites, can cause low adhesion performance at the interface between matrix and fiber. The low interfacial adhesion is associated to the low polarity and chemical affinity between the matrix and the fiber, which causes formation of voids at the interface, and initiation of failures that compromise the mechanical performance of the composites [35].

The characteristics of polymeric composites reinforced with natural fibers immersed in humid environments are influenced by the nature of the fibers and matrix material, relative humidity and manufacturing techniques. The intensity of the water absorption in composite materials (in liquid or vapor phases) depends on several factors, such as temperature, fiber volumetric fraction, reinforcement orientation, natural fiber permeability, porosity, exposed surface area, diffusivity and the existence of surface protection [36, 37].

The absorption of moisture by polymers reinforced by cellulosic fibers occurs because the dissolution of water in the polymer structure, due to the hydrogen bonds between the water and hydrophilic groups present in the components of the composite and, through the micro cracks at the surface of the composite, which are responsible for the transportation and deposition of water [38].

The water absorbed by polymers consists of both free water and bound water [39]. The free water are water molecules with the ability to move independently through the void spaces, while bound water are water molecules that are delimited to polar groups of the polymers [40].

When a polymer composite reinforced with natural fibers is exposed to moisture, free water penetrates and binds with hydrophilic groups of the fiber, establishing intermolecular hydrogen bonds with the fiber and reducing interfacial adhesion between fiber and matrix. The deterioration process occurs with the swelling of the cellulose fibers which promotes an increase in the stress at the interface regions resulting in its embrittlement and thus, leading to formation of micro cracks in the matrix around the fibers that have swollen [41–43]. This promotes capillarity and transport of moisture via micro cracks, causing deterioration of the fibers, which eventually lead to the definitive takeoff between fibers and matrix. After long

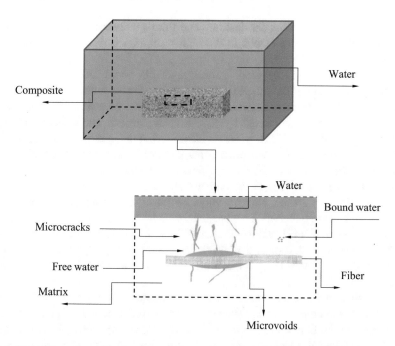

Fig. 1 Water absorption effect in polymer composite reinforced by lignocellulosic fiber

periods, biological activities such as fungal growths eventually degrade the natural fibers [39, 40]. Thus, water absorption is one of the limiting factors that reduces the applicability, and physical and mechanical properties of the composite [44]. Figure 1 shows the moisture conditions (Free water and bound water) and the effect of water on the fiber-matrix interface.

Considering that these materials may be exposed to moisture or even submerged in water during their use, water absorption studies have a constant academic interest, especially when heated at higher temperatures [45, 46].

2.2   Mathematical Modeling for Water Diffusion

Polymeric materials are often exposed to humid environments. The water molecules are able to move inside the polymer and change its physical and mechanical properties. The main parameters determining the moisture sorption mechanism are the chemical composition and the microstructure of the polymers.

Moisture diffusion in polymeric composites is governed by three different mechanisms. The first involves the diffusion of water molecules within micro gaps between polymer chains. The second involves moisture transfer through gaps and faults at the interface between fiber and matrix. This is a result of poor wetting and impregnation during initial fabrication. The third involves the transport of water molecules through micro cracks in the matrix originated from the manufacturing process [45, 47, 48].

The high content of cellulose in the vegetable fibers favors even more the penetration of water at the interface through micro cracks induced by the swelling of the fibers, thus, leading to the insufficiency of the composite [50]. Because the cracks in the composite caused by increased dimensions, the moisture transport increases inside the solid. The water molecules act at the interface, causing the fiber and matrix to detach. Higher temperatures seem to accelerate moisture absorption behavior. When the immersion temperature is increased, the moisture saturation time is greatly reduced due to the higher water diffusion rates.

Due to the importance, several works on water absorption in fiber-reinforced polymer composite materials have been reported in the literature: Espert et al. [45], Pothan and Thomas [46], Bismarck et al. [49], Nair and Thomas [50], Santos et al. [51], Chow et al. [52], Osman et al. [53], Cavalcanti et al. [54], Nóbrega et al. [55], Badia et al. [56], Melo [57], Huner [58], Fonseca et al. [59], Fuentes et al. [60], Bezerra et al. [61], Liu et al. [62].

The diffusion process of water in polymeric composites is very slow. Generally, the process time until the composite reaches saturation conduction is a few months, depending on the fiber content, size and shape of the composite and environmental conditions, especially temperature. In the case of water vapor absorption, the relative humidity also plays an important role. Thus, in order to obtain faster information and with lower process costs, researchers have studied the water absorption process using numerical simulation, that is, theoretically, from developed mathematical models.

Different models and their analytical and numerical solutions have been suggested to describe the kinetics of water sorption in polymer composites [63]. However, these works consider the diffusion through the solid to be unidirectional [64–69]. When dealing with three-dimensional problems, there are still few scientific papers related [4, 9, 57, 70].

Among the different models proposed in the literature to describe the water sorption kinetics in polymer composites, we can cite the Jacob-Jones's model, Fick's model, model with variable diffusivity, and the Langmuir's model. The most commonly used is the Fick's model, which states that water migrates inside the solid purely by diffusion [71, 72]. Other models also take into account chemical reactions and leaching of low molecular weight components that affect the kinetics of water sorption. These models are described in the next section.

2.2.1 The Fick's Model

In this model, it is assumed that the moisture sorption occurs only by diffusion. According to Fick's first law, the diffusive flux (J) is directly proportional to the concentration gradient. Then, we can write:

$$J = -D\frac{\partial M}{\partial y} \tag{1}$$

The Fick's second Law has been widely used, since it establishes the moisture diffusion in terms of concentration gradient in the solid as follows [73]:

$$\frac{\partial M}{\partial t} = \nabla \cdot (D\nabla M) \tag{2}$$

where $\frac{\partial M}{\partial t}$ is the moisture transfer rate and D is the diffusion coefficient. In general the diffusion coefficient is considered constant or depending on the temperature and/or moisture content.

Considering equilibrium boundary condition at the surface of the solid, the exact solution from Eq. (2) as applied for a flat plate generates a mathematical expression for the average moisture content of the sample as follows:

$$\overline{M}(t) = M_\infty - 2\frac{(M_\infty - M_0)}{\pi^2}\sum_{k=1}^{\infty}\frac{\left[1 - (-1)^k\right]^2}{k^2}\exp\left[-\left(\frac{\pi k}{2a}\right)^2 Dt\right] \tag{3}$$

In this expression, M_0 and M_∞ represent the initial and the equilibrium moisture content of the sample, respectively, 2a is the thickness of the sample, and t is the time. The proposed model contains as parameters: the mass diffusivity and the equilibrium moisture content. According to Crank [71], if the sorption curve is drawn on a diagram whose abscissa axis is $\sqrt{t}$ and using experimental data, the mass diffusivity can be determined from the slope of the curve as follows

$$D = \frac{\pi a^2}{16t}\left(\overline{M}^*\right)^2 \tag{4}$$

where $\overline{M}^* = \frac{\overline{M}(t) - M_0}{M_\infty - M_0}$ is the dimensionless average moisture content of the sample at time t.

2.2.2 Jacob's-Jones Model

In the Jacob's-Jones model, it is assumed that the material consists of two phases of different densities and consequently different sorption properties [63, 74]. It is considered that the moisture sorption process in both phases obeys Fick's law. In this model, the formation of chemical bonds between the water molecules and the polymer is neglected. Further, the model admits that in each phase, the water diffusion process is controlled only by the density of that phase. Thus, for a flat plate and considering equilibrium boundary condition, we obtain the exact solution of Eq. (2). From the exact solution is possible to calculate the average moisture content in each phase of the material is expressed below:

$$\overline{M}_1(t) = M_{\infty 1} - 2\frac{(M_{\infty 1} - M_0)}{\pi^2}\sum_{k=1}^{\infty}\frac{\left[1 - (-1)^k\right]^2}{k^2}\exp\left[-\left(\frac{\pi k}{2a}\right)^2 D_1 t\right] \quad (5)$$

$$\overline{M}_2(t) = M_{\infty 2} - 2\frac{(M_{\infty 2} - M_0)}{\pi^2}\sum_{k=1}^{\infty}\frac{\left[1 - (-1)^k\right]^2}{k^2}\exp\left[-\left(\frac{\pi k}{2a}\right)^2 D_2 t\right] \quad (6)$$

Equations (5) and (6), contain four unknown parameters, namely the equilibrium moisture content and the mass diffusivity of each of the phases. As a result, the total average moisture content in the sample is given as follows:

$$\overline{M}(t) = \overline{M}_1(t) + \overline{M}_2(t) \quad (7)$$

According to literature [75], the mass diffusion coefficient of the composite can be calculated from the mass diffusivity of the matrix, the mass diffusivity of the fibers and the volume fraction of the fibers. An epoxy resin, for example, can be considered as a two phase structure of material. The Jacob's-Jones model does not take into account any changes in the microstructure of the material during the sorption process, which may make it difficult to describe the sorption curves, especially in high relative humidity condition, when water vapor sorption occurs. However, it can be used to describe the sorption characteristics of materials with a non-uniform structure. In this model, it is considered that $M_{\infty 1} = M_{\infty 2}$.

2.2.3 Time-Variable Mass Diffusivity Model

According to this model, due to the physical processes occurring in the material (mainly plastification and aging), mass diffusivity decreases with time in proportion to its initial value as follows:

$$\frac{dD}{dt} = -\gamma D(t) \quad (8)$$

The exact solution of Eq. (8) has the following form:

$$D = D_0 e^{-\gamma t} \quad (9)$$

According to Andrikson et al. [75], the model with variable mass diffusivity has as parameters, the mass diffusivity at the initial time D_0, the moisture content in the equilibrium condition M_∞, and the coefficient γ which describes the mass diffusivity variation rate.

For a flat plate under equilibrium boundary condition, the average moisture content of the material can be calculated as follows:

$$\overline{M} = M_\infty - 2\frac{(M_\infty - M_0)}{\pi^2} \sum_{k=1}^{\infty} \frac{\left[1 - (-1)^k\right]^2}{k^2} e^{-\lambda_k^2 F} \tag{10}$$

where $F = \frac{D_0}{\gamma}[1 - \exp(-\gamma t)]$ is the Fourier's criteria, $\lambda_k = \frac{\pi k}{2a}$, $\gamma = \frac{1}{\tau}$, being τ the characteristic time of relaxation.

2.2.4 The Langmuir Model

In the Langmuir model [76], moisture sorption can be explained by assuming that water exists in two phases, the free and bound. In this model, the water molecules of the free phase are adsorbed (becoming bound) with a probability α in a unit time. The water molecules can leave the connected state with a probability β in a unit time. Thus, the diffusion process is described by the same diffusion equation, which is only modified to take into account the two phases of the moisture in the material. According to this model, the average moisture content of a material, as a function of time $\overline{M}(t)$, depends on four parameters: the mass diffusion coefficient D, the equilibrium moisture content M_∞ and the probabilities α and β. For a flat plate under equilibrium boundary condition this parameter will be expressed as follows:

$$\overline{M}(t) = M_\infty \left\{ \frac{\beta}{\alpha + \beta} \exp(-\alpha t) \left[1 - \frac{2}{\pi^2} \sum_{k=0}^{\infty} \frac{\left[1 - (-1)^k\right]^2}{k^2} \exp\left[-\left(\frac{\pi k}{2a}\right)^2 Dt \right] \right] \right.$$
$$\left. + \frac{\beta}{\alpha + \beta} [\exp(-\beta t) - \exp(-\alpha t)] + [1 - \exp(-\beta t)] \right\} \tag{11}$$

Equation (11) is valid when $2\alpha, 2\beta \ll \pi^2 D/(2a)^2$. The values of the parameters α and β are calculated through graphs as a function of the moisture content, which point to the possibility of increasing the transition of water molecules from the bound state to the free state and the opposite too. This model is described in Glaskova et al. [63], Carter and Kilber [76], Bonniau and Bunsell [77], Cotinaud et al. [78] and Apicella et al. [79].

The sorption models considered reflect the moisture sorption process in different ways, considering certain additional processes that occur in the material. Thus, each specific case must be conducted not only with the results of a good approximation between the experimental data and those predicted by the equations, but also by the physical interpretation of the water sorption process included in the models.

This model will be discussed in details in the next section. We notice that if $\alpha = 0$ and $M_0 = 0$, Eq. (11) becomes the results presented to simple diffusion theory (Fick's law).

3 Advanced Langmuir Model

The effects in physical and chemical properties of composite materials induced by the absorption of moisture have received wide attention, not only due to the durability of these systems in operation, but also because of its wide field of application.

The properties of the composite materials depend on the behavior of the matrix, the reinforcement and the fiber/matrix interface. The advancement of the water molecules in structure of the material promotes degradation of the interface and the swelling of the matrix, generating internal stresses and decrease in the mechanical properties. Then, moisture diffusion in composite materials is intrinsically related to the process of degradation of the material, becoming essential to know the water absorption rate, in order to predict its long-term behavior [77].

To describe the water absorption kinetics, the classic Fick treatment is widely used; however, some materials exhibit more complex water absorption kinetics. In some cases, water absorption follows Fick's law, that is, the diffusion process is driven by the water concentration gradient between the medium and the material and continues until reaching hygroscopic equilibrium condition. In others, this model is not applicable [80, 81], because the material presents an anomalous water diffusion, which implies in numerous different phases of sorption that lead to final mass balance. This type of diffusion cannot be easily described by single-phase models, requiring the use of other models that describe the behavior of the material under these conditions [82].

3.1 The Physical Problem

For the physical model was considered a porous plate with thickness 2a, immersed in a fluid solution (water), contained in a thick container (2L + 2a) (Fig. 2). The moisture absorption was analysed in the thickness and the following assumptions were used: The material is considered homogeneous and isotropic; the mass diffusion coefficient is considered constant; the solid is considered axisymmetrical; the process is transient; there is not variation in the dimensions of the material during the diffusion process; the capillary transport through the solid is considered negligible; mass generation inside the solid is neglected; the solid is totally considered dry at the beginning of the process and finally, the solid is in equilibrium with the surrounding at the surface (equilibrium boundary condition).

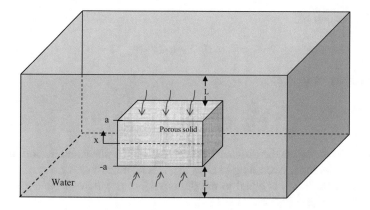

Fig. 2 Geometrical representation of the physical problem

3.2 The Governing Equation

In the Langmuir-Type model, the anomalous moisture absorption can be explained quantitatively by assuming that absorbed moisture consists of two phases, one mobile phase and another bound phase [76]. The model considers the interaction between the polar molecules and the resin molecular groups, predicting the existence of free and bound molecules within the polymer network. This occurs, adding a new parameter to the classical Fick's equation [83]. Considering the assumptions above, the Langmuir equation written in Cartesian coordinates in an one-dimensional approach, is described as:

$$\frac{\partial C}{\partial t} = D\frac{\partial^2 C}{\partial x^2} - \frac{\partial S}{\partial t} \tag{12}$$

where,

$$\frac{\partial S}{\partial t} = \lambda C - \mu S \tag{13}$$

In Equation (13), λ is the probability of a free molecules of water become bound, μ is the probability of a bound molecule become free, D is the mass diffusion coefficient, C is the concentration of free molecules and S is the concentration the bound molecules inside the material. For the proposed problem were considered the following initial and boundary conditions:

- Initial condition:

$$S = C = 0, \quad -a < x < a, \quad t = 0 \tag{14}$$

- Boundary condition:

$$L\frac{\partial C}{\partial t} = -D\frac{\partial C}{\partial x}, \quad x = \pm a, \quad t > 0 \tag{15}$$

where L represents the distance between the solid surface and the top or bottom of the water tank. According to Eq. (15), it is assumed that the rate of solute that leaves the solution is equal to the diffusive flux of solute at the surface of the plane sheet (see Fig. 2).

3.3 Solution Techniques

There is different methods for solution of the diffusion equation such as method of reflection and superposition, method of separation of variables, method of the Laplace transform which are exact method and finite-difference, finite-volume, finite-element methods, etc., which are namely as numerical methods. The choice of a particular technique is related to the easier procedure for a particular physical situation.

3.3.1 Analytical Solution

Based on the works of Carter and Kibler [76], Crank [71] and Santos et al. [81] presents the exact solution for Eqs. (12) and (13) using the method of Laplace Transform. The application of the Laplace Transform consists in convert a partial differential equation in an ordinary differential equation which can be solved more easily. After this procedure, it is calculated the inverse transform to get the original function of the problem [84, 85].

Thus, applying this method in Eqs. (12), (13) and (15), we obtain the following equations:

$$p\overline{C} = -p\overline{S} + D\frac{\partial^2 \overline{C}}{\partial x^2} \tag{16}$$

$$p\overline{S} = \lambda\overline{C} - \mu\overline{S} \tag{17}$$

$$-LC_0 + Lp\overline{C} = -D\frac{\partial\overline{C}}{\partial x}, \quad x = a \tag{18}$$

where $\overline{C}$ and $\overline{S}$ are the Laplace transform of the C and S parameters, respectively. From Eqs. (16) and (17) we obtain $\overline{S}$, as follows:

$$\overline{S} = -\overline{C} + \frac{D}{p}\frac{\partial^2\overline{C}}{\partial x^2} \tag{19}$$

or,

$$\overline{S} = \left(\frac{\lambda}{p+\mu}\right)\overline{C} \tag{20}$$

From Eqs. (19) and (20) we obtain the following equation for $\overline{C}$:

$$\frac{\partial^2\overline{C}}{\partial x^2} + k^2\overline{C} = 0 \tag{21}$$

where,

$$k^2 = -\frac{p}{D}\left(\frac{p+\mu+\lambda}{p+\mu}\right) \tag{22}$$

Since $\overline{C}$ is a function of spatial coordinate x only, we can write Eq. (21), as follows:

$$\frac{d^2\overline{C}}{dx^2} + k^2\overline{C} = 0 \tag{23}$$

The solution of Eq. (23) that gives $\overline{C}$ an even function of x is as follows:

$$\overline{C} = F(p)\cos(kx) \tag{24}$$

The function F(p) is determined by the boundary condition. The value of this function will be:

$$F(p) = \frac{LC_0}{pL\cos(ka) - kD\sin(ka)} \tag{25}$$

So,

$$\overline{C} = \frac{LC_0 \cos(kx)}{pL \cos(ka) - kD \sin(ka)} \tag{26}$$

The inverse Laplace transform of Eq. (26) is not trivial. Herein, we use the partial functions technique like proposed in the literature [71].

Finally, performing all the mathematical formalism proposed by Crank [71], we obtain the inverse transform of $\overline{C}$. Considering the boundary conditions, the domain of the function and with the use of necessary simplifications is obtained the final equation for the parameter C that indicates the concentration of free solute inside the solid during the water absorption process. This equation can be written as follows:

$$C(x,t) = \frac{LC_0}{L + (R + 1)a}$$
$$+ \sum_{n=1}^{\infty} \frac{C_0 \cos(k_n x) e^{(p_n t)}}{\cos(k_n a) \left\{ 1 + \left[1 + \frac{\mu\lambda}{(p_n + \mu)^2} \right] \left[\frac{Lp_n^2 a}{2D^2 k_n^2} + \frac{p_n}{2k_n^2 D} + \frac{a}{2L} \right] \right\}} \tag{27}$$

where p_n and k_n are the eigenvalues and C_0 represent the initial concentration of solute.

The expression for $\overline{S}$ can be obtained using Eq. (20) and its inverse Laplace transform as obtained in the way analogous to Eq. (21). The final equation for the parameter S, which represents the concentration of solute entrapped on the solid, is written as follows:

$$S(x,t) = \left(\frac{\lambda}{\mu} \right) \frac{LC_0}{L + (R + 1)a}$$
$$+ \sum_{n=1}^{\infty} \left(\frac{\lambda}{p_n + \mu} \right) \frac{C_0 \cos(k_n x) e^{(p_n t)}}{\cos(k_n a) \left\{ 1 + \left[1 + \frac{\mu\lambda}{(p_n + \mu)^2} \right] \left[\frac{Lp_n^2 a}{2D^2 k_n^2} + \frac{p_n}{2k_n^2 D} + \frac{a}{2L} \right] \right\}} \tag{28}$$

The total moisture content inside the material in a specific position x and instant t is found from the sum of the amount of free solute and the amount of solute entrapped in accordance with the following equation:

$$M = S + C \tag{29}$$

From Eq. (29), the average moisture content of the solid at different moments of the water uptake can be computed as follows:

$$\overline{M} = \frac{1}{V} \int MdV = \frac{1}{2a} \int_{-a}^{a} Mdx \tag{30}$$

where V is the volume of the solid.

From Eq. (29), the average moisture content of the solid at different moments of the water uptake can be computed as follows:

$$\frac{\overline{M}}{M_e} = 1 - \sum_{n=1}^{\infty} \frac{(1 + \alpha)e^{p_n t}}{1 + \left[1 + \frac{\mu\lambda}{(p_n + \mu)^2}\right]\left[\frac{Lp_n^2 a}{2D^2 k_n^2} + \frac{p_n}{2k_n^2 D} + \frac{a}{2L}\right]} \tag{31}$$

where:

$$\alpha = \frac{L}{(R + 1)a} \tag{32}$$

$$M_e = \frac{LC_0}{(1 + \alpha)a} \tag{33}$$

In Eq. (31), $\overline{M}$ corresponds to the total amount of solute, both free and immobilized to diffusion to a given time t, M_e corresponds to the amount of moisture at the final equilibrium state obtained after an infinite time, and $R = \frac{\lambda}{\mu}$. The terms p_n and k_n together form pairs of eigenvalues and aims to refine the approximate calculation and the results, thus, the higher the number of eigenvalues becomes more accurate the analytical results. They correspond to non-zero roots of the following equation:

$$\frac{Lp_n}{D} = k_n \tan(k_n a) \tag{34}$$

where, the values of k_n are given by Eq. (22).

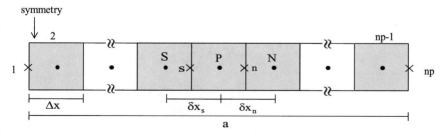

Fig. 3 Representation of the simulation domain with (np-2) control volumes

3.3.2 Numerical Solution

Herein, it was used the finite-volume method for the numerical solution of the governing equation [86]. For the discretization of Eq. (12), the continuous solid of a thickness $\underline{a}$ was considered. Since the solid in question is symmetrical with respect to the center, it was subdivided in (np-1) control volumes, as shown in Fig. 3.

In Fig. 3 each control volume has thickness Δx; S, P and N represent nodal points, while s and n represent the left and right faces of the control volume P, respectively; δx_s and δx_n represent the distance between nodal point P and nodal points S and N, respectively.

- Concentration of free solute:

  (a) Internal points

  The discretization process is done by applying the integral in all terms of Eq. (12) in volume and time, as follows:

$$\int_V \int_t \frac{\partial C}{\partial t} dt dV = \int_t \int_V \frac{\partial}{\partial x}\left(D^C \frac{\partial C}{\partial x}\right) dV dt + \int_t \int_V S^C dV dt \qquad (35)$$

For an one-dimensional problem, we have that $dV = dx$. Thus, after the integration process and using fully implicit formulation, rearranging the terms in the linearized discrete algebraic form applied to the point P, we obtain the following equation:

$$A_P C_P = A_N C_N + A_S C_S + A_P^o C_P^o + B \qquad (36)$$

where,

$$A_P = \left(\frac{\Delta x}{\Delta t} + \frac{D_s^C}{\delta x_s} + \frac{D_n^C}{\delta x_n}\right) \qquad (37)$$

$$A_N = \left(\frac{D_n^C}{\delta x_n}\right) \qquad (38)$$

$$A_S = \left(\frac{D_s^C}{\delta x_s}\right) \qquad (39)$$

$$A_P^o = \left(\frac{\Delta x}{\Delta t}\right) \qquad (40)$$

$$B = S^C \Delta x = \left(\frac{S_P^o - S_P}{\Delta t}\right) \Delta x \qquad (41)$$

The coefficients A_P, A_N and A_S represent the conductance between the nodal points P and their corresponding neighbors. The term A_P^o represents the influence of the value of C^o on the value of C at the current time. Equation (37) is valid for all internal points of the domain except for the boundary and symmetry points.

(b) Symmetry points

The discretized equation for the points of symmetry is obtained in a way analogous to Eq. (37). There is no flux at the center of the solid due to symmetry, thus: $C_P = C_S$ (see Fig. 3). Then, rearranging the terms, we obtain the discretized equation for the points of symmetry as follows:

$$A_P C_P = A_N C_N + A_P^o C_P^o + B \tag{42}$$

where,

$$A_P = \left(\frac{\Delta x}{\Delta t} + \frac{D_n^C}{\delta x_n} \right) \tag{43}$$

$$A_N = \left(\frac{D_n^C}{\delta x_n} \right) \tag{44}$$

$$A_P^o = \left(\frac{\Delta x}{\Delta t} \right) \tag{45}$$

$$B = S^C \Delta x = \left(\frac{S_P^o - S_P}{\Delta t} \right) \Delta x \tag{46}$$

(c) Boundary points

In this case, there is a boundary flux (C'') that must be replaced according to the existing boundary condition, as described by Eq. (15). In this way, the discretized equation for the boundary points is as follows:

$$\left(C_P - C_P^o \right) \frac{\Delta x}{\Delta t} = C'' - D^C \left(\frac{C_P - C_S}{\delta x_s} \right) + S^C \Delta x \tag{47}$$

Equation (4) written as a function of the variable C is represented as follows:

$$L \frac{\partial C}{\partial t} = -D^C \frac{\partial C}{\partial x} \Big|_n \tag{48}$$

For the boundary flux, Eq. (49) is discretized, obtaining the following expression:

$$L\left(\frac{C_n - C_n^o}{\Delta t}\right) = -D_n^C\left(\frac{C_n - C_P}{\partial x_n}\right) \tag{49}$$

Thus, we have:

$$C_n = \left(\frac{\frac{L}{\Delta t}C_n^o + \frac{\Gamma_n^C}{\delta x_n}C_P}{\frac{L}{\Delta t} + \frac{\Gamma_n^C}{\delta x_n}}\right) \tag{50}$$

Since that $L\frac{\partial C}{\partial t} = -D^C\frac{\partial C}{\partial x} = C''$ the expression for the boundary flux is as follows:

$$C'' = \frac{\frac{D_n^C}{\delta x_n}\left(C_P - C_n^o\right)}{1 + \frac{D_n^C \Delta t}{\delta x_n L}} \tag{51}$$

Then, replacing the expression (51) into the (47) and rearranging the terms, the following equation for the boundary points is obtained:

$$A_P C_P = A_S C_S + A_n^o C_n^o + A_P^o C_P^o + B \tag{52}$$

where,

$$A_P = \left(\frac{\Delta x}{\Delta t} + \frac{1}{\frac{\delta x_n}{D_n^C} + \frac{\Delta t}{L}} + \frac{D_s^C}{\delta x_s}\right) \tag{53}$$

$$A_S = \left(\frac{D_s^C}{\delta x_s}\right) \tag{54}$$

$$A_n^o = \left(\frac{1}{\frac{\delta x_n}{D_n^C} + \frac{\Delta t}{L}}\right) \tag{55}$$

$$A_P^o = \left(\frac{\Delta x}{\Delta t}\right) \tag{56}$$

$$B = S^C \Delta x = \left(\frac{S_P^o - S_P}{\Delta t}\right)\Delta x \tag{57}$$

Equations (36), (42) and (52) applied at each control volumes form a system of equations whose solution indicates the concentration of free solute within the solid during the water absorption process.

- Concentration of solute entrapped:
For obtain the equations related to the concentration of solute entrapped in the solid, Eq. (13) in integrated in volume and time as follows:

$$\int_x \int_t \frac{\partial S}{\partial t} \, dt dx = \int_x \int_t (\lambda C - \mu S) \tag{58}$$

resulting in,

$$\left(S_P - S_P^o\right) \frac{\Delta x}{\Delta t} = (\lambda C_P - \mu S_P) \Delta x \tag{59}$$

Reorganized the terms of Eq. (59), we obtain the following equation, valid for all points of the control volume:

$$A_P S_P = A_P^o S_P^o + B \tag{60}$$

where,

$$A_P = \left(\frac{\Delta x}{\Delta t} + \mu \Delta x\right) \tag{61}$$

$$A_P^o = \left(\frac{\Delta x}{\Delta t}\right) \tag{62}$$

$$B = \lambda \Delta x C_P \tag{63}$$

- Total Moisture Content
The total moisture inside the material in a specific position x and instant t is found from the sum of the amounts of free and entrapped solute in accordance as specified in (29):

$$M = S + C \tag{64}$$

- Average Moisture Content
The average moisture content of the solid at any instant of time is given by Eq. (30). This equation in the discretized form is written as follows:

$$\overline{M} = \frac{1}{a} \sum_{i=2}^{np-1} M_i \Delta x_i \tag{65}$$

where np represents the total number of nodal points.

The equation system were solved iteratively using the Gauss-Seidel method, where it was assumed that the numerical solution converged when, from the initial condition, the following convergence criterion was satisfied, at each point of the domain, at a certain time:

$$\left| C_P^{n+1} - C_P^n \right| \leq 10^{-10} \tag{66}$$

where n represents the nth iteration at each instant. For obtain the predicted results, a time step and mesh refinement study was done. After this process we choice a grid with 20 nodal points and a time step $\Delta t = 20\,s$.

4 Application

4.1 Experimental Data

Nóbrega [9] and Nóbrega et al. [55] realized experiments on the water absorption of Caroá fiber-reinforced unsaturated polyester composites. The composite plates were obtained by using the hand layup manufacturing technique. Composites samples of $20 \times 20 \times 3\ mm^3$ were cut-off from these plates, and their edges sealed with resin prior to the water absorption test (to avoid water transport by capillarity) and dried in an air-circulating oven at 105 °C (up to constant weight or dry mass).

The water uptake experiments were carried out according to the following procedure. Firstly, the pre-dried composites samples were immersed fully into a water baths (Fig. 4) kept at 25 °C. At regular intervals the samples were removed from the water bath, wiped with tissue paper to remove surface water and immediately weighted in a electronic balance with precision ±1 mg. Following, the samples were re-immersed in the water bath to continue the sorption process until

Fig. 4 **a** Caroá composite samples and **b** Caroá composites samples in water bath

the equilibrium condition is reached. Each measurement took less than 1 min, so water evaporation at the surface was insignificant.

The results of absorbed moisture were presented as mass of absorbed water by unit dry composites mass. The moisture content was computed as follows:

$$\overline{M}(t) = \frac{W_t - W_0}{W_0} \times 100\% \tag{67}$$

where W_0 and W_t represent the dry weight of the composites samples ($t = 0$) and the wet weight at any specific time t, respectively. Saturation (equilibrium) condition was assumed when the daily weight gain of the composite samples was less than 0.1%.

4.2 Theoretical Data

The formulation was applied to the moisture diffusion process in polymeric composites reinforced with Caroá fibers. The composite studied have very large width and length compared to the thickness. Then, the pure water penetrates only in the direction of the thickness.

For the validation of the model, the numerical result of the average moisture content was compared with analytical results reported by Santos et al. [81] and the experimental data reported by Silva [87] for polymer composite materials reinforced by Caroá fiber (T = 25 °C). Table 1 presents the geometric parameters used in the simulation.

Santos et al. [81] compared the analytical and experimental data of the average moisture content along the transient process at the Temperature T = 25 °C. From this comparison, it was estimated the mass diffusion coefficient and the probabilities μ and λ, of the model, using the least squares error technique, given by the following equation:

$$ERQM = \sum_{i=1}^{n} \left[\overline{M}_{predicted} - \overline{M}_{experimental} \right]^2 \tag{68}$$

where n is the number of experimental points. According to the authors, an average quadratic error of 0.047467 $kg_{water}/kg_{dry\ solid})$ was obtained. The estimated data are shown in Table 2.

Table 1 Geometrical parameters used in the simulation

Parameter	Value
L (m)	0.3
a (m)	1.5×10^{-3}

Table 2 Physical parameters used in the simulation

Parameter	Value
D (m^2 s^{-1})	7.020×10^{-12}
μ (s^{-1})	1.697×10^{-6}
λ (s^{-1})	0.836×10^{-6}

Table 3 Values of the some p and k eigenvalues

n	k_n (m^{-1})	P_n (m^{-1})
1	1049.07	-7.5768×10^{-6}
2	1058.21	-1.29658×10^{-6}
3	3142.29	-6.06761×10^{-6}
4	3171.09	-1.44941×10^{-6}
5	5236.41	-1.67169×10^{-6}

From these values, they were found 30 pair of eigenvalues more one pair of imaginary root to form the exact solution of Eqs. (27), (28) and (31). Table 3 presents five values of p_n and k_n determined in this work.

The predicted moisture content inside the composite materials along the time is particularly important because, from the results obtained, it is possible to predict what areas most susceptible to stress causing fissures and deformations, thereby decreasing the quality of the product [86].

Figure 5 illustrates the comparison between numerical, analytical and experimental data of the average moisture content obtained during water absorption in Caroá fiber reinforced polymer composites at 25 °C.

Fig. 5 Comparison among the numerical, analytical and experimental dimensionless average moisture content of polymer composites reinforced with Caroá fiber during the water absorption process (T = 25 °C)

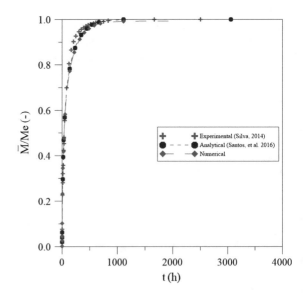

The numerical results depend strongly on the boundary conditions, thermo-physical properties and the geometry considered. The observed discrepancies between experimental and numerical data can be attributed to the lack of suitable boundary conditions for the model and the consideration of constant properties. However, from the analysis of the graph it can be observed that the numerical results showed a good agreement between the predict and analytical and experimental results, which shows that the mathematical model presents an adequate description of the diffusion process inside the material.

In Fig. 6, new results of the average moisture content are plotted as a function of dimensionless parameter $F_o = \dfrac{Dt}{(R+1)a^2}$ namely modified Fourier number of mass transfer aiming a more general mathematical analysis. This way the results are independent of mass diffusivity, λ and μ probabilities, dimensions of the sheet and time. Based on the graph analysis it is observed that the water sorption is very quick in the early stages up to a certain modified Fourier number of mass transfer approximately 2.0, and tends to decline for long exposure times until to achieve equilibrium point (saturation condition $M_e = 14.488\%$), where the dimensionless average moisture content tends to 1. The time to achieve the hygroscopic equilibrium was estimated to $F_o \cong 10$, $F_o \approx 12$. The graph related to the kinetics of the problem has characteristics compatible with results found in literature [38, 64].

Figure 7 shows the graph of the variation of free solute concentration (analytical and numerical) along the thickness of the solid, given by the C/Ce, where Ce (estimated as Ce = 0.09778) represents the equilibrium concentration of the solute in the fluid medium. Based on the analysis of this figure it is observed that for shorter times the concentration variation is higher close to the surface of the material, that is, there is a high concentration gradient of free water, in these

Fig. 6 Dimensionless average moisture content as a function of modified Fourier number of mass transfer (T = 25 °C)

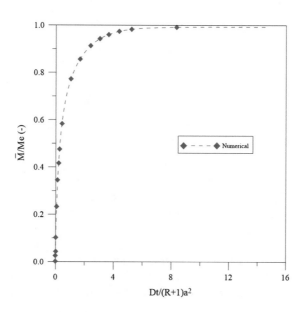

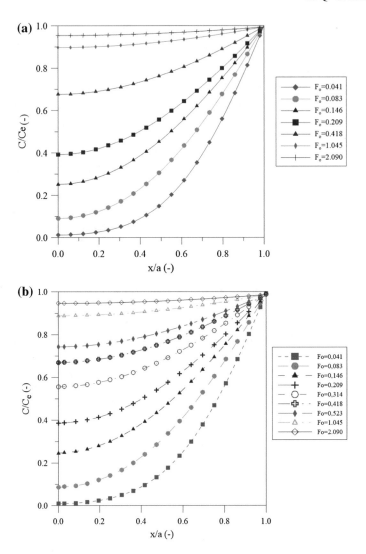

Fig. 7 Free solute concentration distribution along the material for different instants (in terms of modified Fourier number of mass transfer). **a** Analytical and **b** Numerical results

regions. With increasing time, this relationship tends to approach 1, reaching the hygroscopic equilibrium condition (saturation) due to higher absorption of water. Beside, at any point within the solid, the moisture content increases with time until to reach a state of equilibrium.

Figure 8 shows the graph of the variation of bound (entrapped) water molecules (numerical) along the solid thickness. Based on the analysis of this graph it is observed that the increase in the content of bound molecules depends on the increase in the concentration of free molecules into the material. The major water

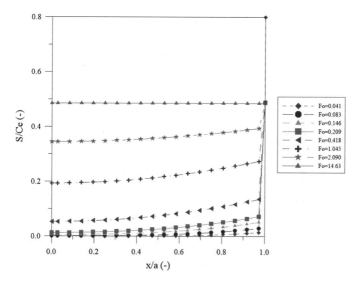

Fig. 8 Trapped solute concentration distribution along the material for different instants of the water absorption process (numerical results)

concentration gradients are found near the solid surface. For longer times, or when there is a greater amount of free molecules within the material we have greater the number of trapped molecules and this occurs until equilibrium. This condition occurs as $\frac{\partial S}{\partial t} = 0$, or yet, $\lambda C = \mu S$, which implies in: $\frac{S}{C_e} = \left(\frac{\lambda}{\mu}\right)\frac{C}{C_e} = \frac{\lambda}{\mu}$, for $t \rightarrow \infty$.

The observed behaviour at the beginning of absorption process is particularly Fickian, that is, moisture migration of water molecules in "free" state, however, as more moisture is absorbed, the water diffusion rate decreases. This fact can be explained by two phenomena: (a) as the more water absorbed more molecules are linked to the polymer chains, thus reducing the amount of water that can be absorbed and (b) the relaxation rate becomes larger than the diffusion rate, controlling the final stages of the process [88].

Figure 9 shows the graph of the total moisture present into the material, obtained by the sum of the number of free molecules to be diffused and the number of molecules trapped along its thickness. In regions near the surface, water absorption is faster, because exist a larger area in direct contact with water. The water penetrates the interior of the material generating a higher concentration gradient along the thickness and decreasing with the increase of the immersion time. Thus, at any point inside the solid, the moisture content increases with time until it reaches its equilibrium condition, i.e. its saturation point. It is intuitive to say that with a longer immersion time there is an increase in the amount of molecules trapped within the material while decreasing the amount of free molecules to diffuse.

As final comment we notice that water absorption is facilitated when polymer molecules have clusters capable of forming hydrogen bonds. Plant fibers are rich in

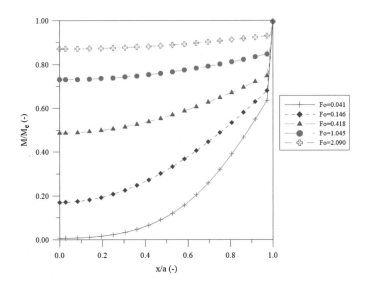

Fig. 9 Moisture content profile inside the porous solid for different instant of the water absorption process (numerical results)

cellulose, hemicellulose and lignin which have hydroxy groups, i.e., have high affinity for water. The absorption of water by the resin, in turn, can be considered practically null, since it presents a considerable hydrophobic character [89]. Further, the addition of the plant fibers to the resin generates an increase in the water absorption levels, so an important parameter to be analyzed is how much water is being absorbed by the material over time.

The effects caused due to long time exposure to moisture may be irreversible due to the water molecules affinity with specific functional groups of the polymeric matrices. The destructive changes usually occurs due to degradation of probable existing physical-chemical interactions between the resin and fiber, as a consequence, there is a shift in the fiber, causing delamination and reduction in the composite material properties. Thus, understanding this process is crucial to predict the quality of the material under wet environments.

From the physical and mathematical viewpoints is important to analyse quantitatively what happens with the general solution for the extreme values of the probability μ which correspond to very fast and very slow process. When μ is very large as compared with λ, the process is very rapid compared with diffusion. Then, the immobilized component is in equilibrium condition with the component free to diffuse into the sheet, thus, the process is controlled by diffusion. However, if $\mu \rightarrow 0$ so, the process is infinitely slow, the plane sheet takes up, by simple diffusion only the fraction of solute which the sheet can accommodate in the freely diffusing state and none in the immobilized state. Besides, for the case where D is very large, the diffusion is so rapid that the concentration of free and immobilized solute are almost uniform through the sheet during the water uptake process [85].

5 Concluding Remarks

In this chapter, an analysis of anomalous moisture sorption in vegetable fiber-reinforced composites has been carried out. An advanced Fick's model namely Langmuir-Type model has been used for prediction of water uptake considering a transient and one-dimensional approach.

The mathematical model used describes properly the water diffusion inside the polymer composites reinforced with Caroá fiber, since the both exact and numerical solutions obtained showed good agreement with the experimental and analytical data of the average moisture content along the water absorption process.

Some all results are shown in the dimensionless form aiming a more general mathematical analysis. From the analysis of the predicted results, it was confirmed that the water absorption is fast in the initial stages and tends to decay for long exposure times to water until reaching the equilibrium point (hygroscopic saturation condition). Further, the concentration gradients are higher at the surface of the material and the higher the free solute concentration, the higher the immobilized solute concentration inside the polymer composite.

Acknowledgements The authors thanks to CNPq, CAPES and FINEP (Brazilian Research Agencies) for financial support and the authors of the references cited in text which helps in the improvement.

References

1. Peters, S.T.: Handbook of Composites. Chapman & Hall, Cambridge University Press, Cambridge, England (1999)
2. Shackelford, J.F.: Materials Science for Engineers, 6th edn. Pearson Prentice Hall, New Jersey, USA (2005)
3. Soutis, C.: Fibre reinforced composites in aircraft construction. Prog. Aerosp. Sci. **41**(2), 143–151 (2005)
4. Cavalcanti, W. S.: Composites polyester/woven plant-glass: mechanical characterization and simulation of water sorption, Doctorate Thesis in Process Engineering, University Federal of Campina Grande—UFCG, Campina Grande, PB, Brazil (2006) (in portuguese)
5. Mouzakis, D.E., Zoga, H., Galiotis, C.: Accelerated environmental ageing study of polyester/glass fiber reinforced composites (GFRPCs). Compos. B Eng. **39**(3), 467–475 (2008)
6. Ratna, D.: Handbook of Thermoset Resins, pp. 19–21. Smithers Publishers, Shawbury, United Kingdom (2009)
7. Callister Jr., W.D.: Materials Science and Engineering an Introduction, 8th edn. Wiley, New York, USA (2012)
8. Mallick, P.K.: Fiber-Reinforced Composites: Materials, Manufacturing, and Design, 3rd edn. CRC Press, Taylor and Francis Group, New York (2007)
9. Nóbrega, M.M.S.: Composites of polyester matrix with caroá fiber *Neoglaziovia Variegata*: mechanical characterization and sorption of water. Doctorate Thesis in Process Engineering, Federal University of Campina Grande, Campina Grande, PB, Brazil (2007) (in Portuguese)
10. Jayamol, G., Bhagawan, S.S., Thomas, S.: Effects of environment on the properties of low-density polyethylene composites reinforced with pineapple-leaf fibre. Compos. Sci. Technol. **58**(9), 1471–1485 (1998)

11. Bledzki, A.K., Gassan, J.: Composite reinforced with cellulose based fiber. Prog. Polym. Sci. 24(2), 221–274 (1999)
12. Joseph, K., Medeiros, E.S., Carvalho, L.H.: Composites of polyester matrix reinforced by short fibers of sisal. Polímeros 9(4), 136–141, São Carlos (1999) (in Portuguese)
13. Joseph, S., Sreekala, M.S., Oommen, Z., Thomas, S.: A comparison of the mechanical properties of phenol formaldehyde composites reinforced with banana fibers and glass fibers. Compos. Sci. Technol. 62(14), 1857–1868 (2002)
14. Medeiros, E.S., Agnelli, J.A.M., Joseph, K., Carvalho, L.H., Mattoso, L.H.C.: Curing behavior of a novolac-type phenolic resinanalyzed by differential scanning calorimetry. J. Appl. Polym. Sci. 90(6), 1678–1682 (2003)
15. Nóbrega, M.M.S., Fonseca, V.M.; Carvalho, L.H.: Use of Caroá and macambira fibers in polyester matrix composites. In: 8th Brazilian Polymer Congress, Proceedings, pp. 1231–1232. Águas de Lindóia, Brazil (2005) (in Portuguese)
16. Agarwal, R., Saxena, N.S., Sharma, K.B., Thomas, S., Pothan, L.A.: Thermal conduction and diffusion through polyester composites. Indian J. Pure Appl. Phys. 44(10), 746–750 (2006)
17. Carvalho, L.H., Cavalcanti, W.S.: Mechanical properties of composite polyester/hybrid sisal/glass. Polymers 16(1), 33–37 (2006) (in Portuguese)
18. Carvalho, L.H., Canedo, E.L., Neto, S.F., Lima, A.G.B.: Moisture transport process in vegetable fiber composites: theory and analysis for technological applications. In: Delgado, J. M.P.Q. (ed.) Industrial and technological applications of transport in porous materials, pp. 37–62. Berlin, Heidelberg, Springer (2013)
19. Uday Kiran, C., Ramachandra Reddy, G., Dabade, B.M., Rajesham, S.: Tensile properties of sun hemp, banana and sisal fiber reinforced with polyester composites. J. Reinf. Plast. Compos. 26(10), 1043–1050 (2007)
20. Haneefa, A., Bindu, P., Arvind, I., Thomas, S.: Studies on tensile and flexural properties of short banana/glass hybrid fiber reinforced polystyrene composites. J. Compos. Mater. 42(15), 1471–1489 (2008)
21. Costa, D.S., Peres, M.N.P.B., Barreira, R.M., Silva, R.L.B.,Silva, V.L.D., Sá, F.A., Souza, J. A.S.: Composites reinforced with sisal and mauve fibers: analysis of the tensionxlength fiber. In: 20th Brazilian Congress of Engineering and Science Materials, Joinville, Brazil (2012) (in Portuguese)
22. Venkateshwaran, N., Elayaperumal, A., Sathiya, G.K.: Prediction of tensile properties of hybrid-natural fiber composites. Compos. B Eng. 43(2), 793–796 (2012)
23. Santos, A.C.L., Miranda, C.S., Carvalho, G.G.P., Carvalho, R.F., Fiuza, R.P., Jose, N.M.: Effect of surface treatments on banana fiber from the state of Bahia. In: 20th Brazilian Congress of Engineering and Science Materials, Joinville, Brasil (2002) (in Portuguese)
24. Smith, F.C., Moloney, L.D., Matthews, F.L.: Fabrications of woven fibre/polycarbonate repair patches. Compos. A Appl. Sci. Manuf. 27A(11), 1089–1095 (1996)
25. Idicula, M., Boudenne, A., Umadevi, L., Ibos, L., Candau, Y., Thomas, S.: Thermophysical properties of natural fibre reinforced polyester composites. Compos. Sci. Technol. 66(15), 2719–2725 (2006)
26. Cavalcanti, W.S.: Mechanical properties of polyester/jute composites: effects of surface treatments, thermal aging and water absorption. Master Dissertation in Chemical Engineering, Federal University of Campina Grande, Campina Grande, PB, Brazil (2000) (in Portuguese)
27. Faruk, O., Bledzki, A.K., Fink, H.P., Sain, M.: Biocomposites reinforced with natural fibers: 2000–2010. Prog. Polym. Sci. 37(11), 1552–1596 (2012)
28. Ishizaki, M.H., Visconte, L.L.Y., Furtado, C.R.G., Leite, M.C.A.M., Leblanc, J.L.: Mechanical and morphological characterization of polypropylene and green coconut fiber composites: Influence of fiber content and mixing conditions. Polímeros 16(3), 182–186 (2006) (in Portuguese)
29. D'almeida, A.L.F.S.; Carvalho, L.H.; D'almeida, J.R.M.: Characterization of caroá (Neoglaziovia variegata) fibers. In: World Polymer Congress and 41° International Symposium on Macromolecules. Procedings. Rio de Janeiro, Brazil (2006)

30. Kalia, S., Kaith, B.S., Kaur, I.: Cellulose fibers: bio- and nano-polymer composites green chemistry and technology. Springer-Verlag, Berlin, Germany (2011)
31. Silva, H.S.P.: Development of polymeric composites with curauá fibers and hybrids with glass fibers. Master Dissertation in Materials, Metalutgy and Mining Engineering. Federal University of Rio Grande do Sul, Porto Alegre, Brazil (2010) (in Portuguese)
32. Trindade, W.G., Hoareau, W., Megiatto, J.D., Razera, I.A.T., Castellan, A., Frollini, E.: Thermoset phenolic matrices reinforced with unmodified and surface-grafted furfuryl alcohol sugar cane bagasse and curaua fibers: properties of fibers and composites. Biomacromol 6(5), 2485–2496 (2005)
33. Levy Neto, F., Pardini, L.C.: Structural composites: science and technology. Ed. Edgar Blucher, São Paulo, Brazil (2006) (in Portuguese)
34. Silva, R.V.: Polyurethane resin composite derived from castor oil and vegetable fibers. Doctorate Thesis in Science and Materials Engineering. University of São Paulo, São Carlos, SP, Brazil (2003) (in Portuguese)
35. Paul, S.A., Boudenne, A., Ibos, L., Candau, Y.: Effect of fiber loading and chemical treatments on thermophysical properties of banana fiber/polypropylene commingled composite materials. Compos. A Appl. Sci. Manuf. 39(9), 1582–1588 (2008)
36. Joseph, P.V., Rabello, M.S., Mattoso, L.H., Joseph, K., Thomas, S.: Environmental effects on the degradation behavior of sisal fiber reinforced polypropylene composites. Compos. Sci. Technol. 62(10–11), 1357–1372 (2002)
37. Dhakal, H.N., Zhang, Z.Y., Richardson, M.O.W.: Effect of water absorption on the mechanical properties of hemp fibre reinforced unsaturated polyester composites. Compos. Sci. Technol. 67(7–8), 1674–1683 (2007)
38. Sreekala, M.S., Kumaran, M.G., Thomas, S.S.: Water sorption in oil palm fiber reinforced phenol formaldehyde conposites. Compos. A 33, 763–777 (2002)
39. Chen, H., Miao, M., Ding, X.: Influence of moisture absorption on the interfacial strength of bamboo/vinyl ester composites. Compos. A Appl. Sci. Manuf. 40(12), 2013–2019 (2009)
40. Azwa, Z.N., Yousif, B.F., Manalo, A.C., Karunasena, W.: A review on the degradability of polymeric composites based on natural fibres. Mater. Des. 47, 424–442 (2013)
41. Li, Y., Mai, Y.W., Ye, L.: Sisal fibre and its composites: a review of recent developments. Compos. Sci. Technol. 60(11), 2037–2055 (2000)
42. Wambua, P., Ivens, J., Verpoest, I.: Natural fibres: can they replace glass in fibre reinforced plastics? Compos. Sci. Technol. 63(9), 1259–1264 (2003)
43. Aziz, S.H., Ansell, M.P., Clarke, S.J., Panteny, S.R.: Modified polyester resins for natural fibre composites. Compos. Sci. Technol. 65(3–4), 525–535 (2005)
44. Hodzic, A., Shanks, R.: Natural fibre composites—materials, processes and properties. Ed. Woodhead Publishing Limited, Philadelphia, USA (2014)
45. Espert, A., Vilaplana, F., Karlsson, S.: Comparison of water absorption in natural cellulosic fiber from wood and one-year crops in polypropylene composites and its influence on their mechanical properties. Compos. A Appl. Sci. Manuf. 35(11), 1267–1276 (2004)
46. Pothan, L., Thomas, S.: Effect of hybridization and chemical modification on the water-absorption behaviour of banana fiber-reinforced polyester composites. J. Appl. Polym. Sci. 91(6), 3856–3865 (2004)
47. Lin, Q., Zhou, X., Dai, G.: Effect of hydrothermal environment on moisture absorption and mechanical properties of wood flour-filled polypropylene composites. J. Appl. Polym. Sci. 85 (14), 2824–2832 (2002)
48. Comyn, J.: Polymer permeability, p. 383. Elsevier, London, England (1985)
49. Bismarck, A., Askargorta, I.A., Springer, J., Lampke, T., Wielage, B., Stamboulis, A.: Surface characterization of flax, hemp and cellulose fibres; surface properties and the water uptake behaviour. Polym. Compos. 23(5), 872–894 (2002)
50. Nair, M.K.C., Thomas, S.: Effect of ageing on the mechanical properties of short sisal fibre reinforced polystyrene composites. J. Thermoplast. Compos. Mater. 16(3), 249–271 (2003)

51. Santos, D.G., Lima, A.G.B., Pinto, M.V.S.: Mechanical characterization and water sorption in polyester matrix composites reinforced with sisal fiber: An experimental investigation. Deff. Diff. Forum **369**, 131–134 (2016)
52. Chow, C.P.L., Xing, X.S., Li, R.K.Y.: Moisture absorption studies of sisal fibre reinforced polypropylene composites. Compos. Sci. Technol. **67**(2), 306–313 (2007)
53. Osman, E., Vakhguelt, E., Sbarski, I., Mutasher, S.: Water absorption behavior and its effect on the mechanical properties of kenaf natural fiber unsaturated polyester composites. In: 18th International Conference on Composite Materials, Edinburgh, Scotland (2009)
54. Cavalcanti, W.S., Carvalho, L.H., Lima, A.G.B.: Sorption of water in polyester-unsaturated composites reinforced with jute and jute-glass fabric: modeling and simulation and experimentation. Polymers, **20**(1), 78–83 (2010) (in Portuguese)
55. Nóbrega, M.M.S., Cavalcanti, W.S., Carvalho, L.H., Lima, A.G.B.: Water absorption in unsaturated polyester composites reinforced with caroá fiber fabrics: modeling and simulation. Materialwiss. Werkstofftech. **41**(5), 300–305 (2010)
56. Badia, J.D., Kittikorn, T., Strömberg, E., Santonja-Blasco, L., Martínez-Felipe, A., Ribes-Greus, A., Ek, M., Karlsson, S.: Water absorption and hydrothermal performance of PHBV/sisal biocomposites. Polym. Degrad. Stab. **108**, 166–174 (2014)
57. Melo, J. B. C. A.: Water absorption in fiber reinforced polymer composites of pineapple leaf: modeling and simulation. Doctorate Thesis in Process Engineering. Federal University of Campina Grande, Campina Grande, Brazil (2014) (in Portuguese)
58. Huner, U.: Effect of water absorption on the mechanical properties of flax fiber reinforced epoxy composites. Adv. Sci. Technol. **9**(26), 1–6 (2015)
59. Fonseca, A.A.P., Arellano, M., Rodrigue, D., Nunez, R.G., Ortiz, J.R.R.: Effect of coupling agent content and water absorption on the mechanical properties of coir-agave fibers reinforced polyethylene hybrid composites. Polym. Compos. **37**(10), 3015–3024 (2016)
60. Fuentes, C.A., Ting, K.W., Dupont-Gillain, C., Steensma, M., Talma, A.G., Zuijderduin, R., Van Vuure, A.W.: Effect of humidity during manufacturing on the interfacial strength of non-pre-dried flax fibre/unsaturated polyester composites. Compos. A Appl. Sci. Manuf. **84**, 209–215 (2016)
61. Bezerra, A.F.C., Cavalcanti, W.S., Lima, A.G.B., Souza, M.J.: Unsaturated polyester composite/caroá Fiber (*Neoglaziovia Variegata*): Water sorption and mechanical properties. In: 3nd Brazilian Conference on Composite Materials—BCCM-3, Gramado, Brazil (2016)
62. Liu, W., Qiu, R., Li, K.: Effects of fiber extraction, morphology, and surface modification on the mechanical properties and water absorption of bamboo fibers-unsaturated polyester composites. Polym. Compos. **37**(5), 1612–1619 (2016)
63. Glaskova, T.I., Guedes, R.M., Morais, J.J., Aniskevich, A.N.: A comparative analysis of moisture transport models as applied to an epoxy binder. Mech. Compos. Mater. **43**(4), 377–388 (2007)
64. Choi, H.S., Ahn, K.J., Nam, J.D., Chun, H.J.: Hygroscopic aspects of epoxy/carbon fiber composite laminates in aircraft environments. Compos. A Appl. Sci. Manuf. **32**(5), 709–720 (2001)
65. Srihari, S., Revathi, A., Rao, R.M.V.G.K.: Hygrothermal effects on RT-cured glass-epoxi composites in immersion environments. Part A: Moisture absorption characteristics. J. Reinf. Plast. Compos. **21**(11), 983–991 (2002)
66. Yao, J., Ziegmann, G.: Water absorption behavior and its influence on properties of GRP pipe. J. Compos. Mater. **41**(8), 993–1008 (2007)
67. Najafi, S.K., Kiaefar, A., Hamidina, E., Tajvidi, M.: Water absorption behavior of composites from sawdust and recycled plastics. J. Reinf. Plast. Compos. **26**(3), 341–348 (2007)
68. Czél, G., Czigány, T.: A study of water absorption and mechanical properties of glass fiber/polyester composite pipes-effects of specimen geometry and preparation. J. Compos. Mater. **42**(26), 2815–2827 (2008)
69. Katzman, H.A., Castaneda, R.M., Lee, H.S.: Moisture diffusion in composite sandwich structures. Compos. A Appl. Sci. Manuf. **39**(5), 887–892 (2008)

70. Cruz, V.C.A.: Composites of polyester matrix with fibers of macambira (bromeliad laciniosa): modeling, simulation and experimentation. Doctorate Thesis in Process Enginerring. Federal University of Campina Grande, Campina Grande, Brazil (2013) (in Portuguese)
71. Crank, J.: The Mathematics of Diffusion, 2nd edn. Oxford University Press, Oxford (1975)
72. Xiao, G.Z., Shanahan, M.E.R.: Swelling of DGEBA/DDA epoxy resin during hygrothermal ageing. Polymer **39**(14), 3253–3260 (1998)
73. Lima, A.G.B.: Phenomenon diffusion in solid spheroidal prolates. Case study: Banana drying. Ph.D. Thesis in Mechanical Engineering. State University of Campinas, Campinas (1999) (in Portuguese)
74. Maggana, C., Pissis, P.: Water sorption and diffusion studies in an epoxy resin system. J. Polym. Sci. Part B-Polym. Phys. **37**(11), 1165–1182 (1999)
75. Andrikson, G.A., Mochalov, V.P., Aniskevich, A.N.: Principle of modified time scale for tasks of nonstationary moisture diffusion in polymer materials. Mech. Compos. Mater. **1**, 153–170 (1980) (in Russian)
76. Carter, H.G., Kibler, K.G.: Langmuir-type model for anomalous moisture diffusion in composite resins. J. Compos. Mater. **12**(2), 118–131 (1978)
77. Bonniau, P., Bunsell, A.R.: A comparative study of water absorption theories applied to glass epoxy composites. J. Compos. Mater. **15**(3), 272–293 (1981)
78. Cotinaud, M., Bonniau, P., Bunsell, A.R.: The effect of water absorption on the electrical properties of glass-fibre reinforced epoxy composites. J. Mater. Sci. **17**(3), 867–877 (1982)
79. Apicella, A., Estiziano, L., Nicolais, L., Tucci, V.: Environmental degradation of the electrical and thermal properties of organic insulating materials. J. Mater. Sci. **23**(2), 729–735 (1988)
80. Grace, L.R., Altan, M.C.: Characterization of anisotropic moisture absorption in polymeric composites using hindered diffusion model. Compos. A Appl. Sci. Manuf. **43**(8), 1187–1196 (2012)
81. Santos, W.R.G., Melo, R.Q.C., Lima, A.G.B.: Water absorption in polymer composites reinforced with vegetable fiber using Langmuir-type model: An exact mathematical treatment. Defect Diffus. Forum **371**, 102–110 (2016)
82. Perreux, D., Suri, C.: A study of the coupling between the phenomena of water absorption and damage in glass/epoxy composite pipes. Compos. Sci. Technol. **57**(9–10), 1403–1413 (1997)
83. Popineau, S., Rondeau-Mouro, C., Sulpice-Gaillet, C., Shanahan, M.E.: Free/bound water absorption in an epoxy adhesive. Polymer **46**(24), 10733–10740 (2005)
84. Fu, Z., Chen, W., Yang, H.: Boundary particle method for Laplace transformed time fractional diffusion equations. J. Comput. Phys. **235**, 52–66 (2013)
85. Zhu, S., Satravaha, P., Lu, X.: Solving linear diffusion equations with the dual reciprocity method in Laplace space. Eng. Anal. Bound. Elem. **13**(1), 1–10 (1994)
86. Melo, R.Q.C., Santos, W.R.G., Lima, A.G.B.: Applying the Lagmuir-type model on the water absorption in vegetable fiber reinforced polymer composites: A finite-volume approach. In: XXXVIII Iberian Latin-American Congress on Computational Methods in Engineering, Florianópolis, Brazil, 5–8 Nov 2017
87. Silva, C.J.: Water absorption in composite materials of vegetal fiber: modeling and simulation via CFX. Master Dissertation in Mechanical Engineering. Federal University of Campina Grande, Brazil (2014) (in Portuguese)
88. Placette, M.D., Fan, X., Zhao J.H., Edwards, D.: A dual stage model of anomalous moisture diffusion and desorption in epoxy mold compounds. In: 12th International Conference on Thermal, Mechanical and Multi-Physics Simulation and Experiments in Microelectronics and Microsystems (EuroSimE). IEEE, Linz, Austria (2011)
89. Sanchez, E.M., Cavani, C.S., Leal, C.V., Sanchez, C.G.: Composites of unsaturated polyester resin with sugarcane bagasse: influence of fiber treatment on properties. Polymer **20**(3), 194–200 (2010)
90. Kumosa, L., Benedikt, B., Armentrout, D., Kumosa, M.: Moisture absorption properties of unidirectional glass/polymer composites used in composite (non-ceramic) insulators. Compos. A Appl. Sci. Manuf. **35**(9), 1049–1063 (2004)

Liquid Injection Molding Process in the Manufacturing of Fibrous Composite Materials: Theory, Advanced Modeling and Engineering Applications

M. J. Nascimento Santos, João M. P. Q. Delgado, Antonio Gilson Barbosa de Lima and I. R. Oliveira

Abstract The purpose of this chapter is to provide theoretical and experimental information about polymer composite manufacturing reinforced with fiber by using Resin Transfer Molding process. It is a process, in which the liquid resin is injected in a closed mold with a fibrous preform inserted. This physical process is similar to the fluid flow in porous media, thus, the process control becomes essential. Here, diverse topics related to this theme, such as, theory, experiments, advanced macroscopic mathematical modeling, in which is included the effect of the resin sorption by fibers, exact solution of the governing equations, and technological applications are presented and well discussed. The study clarifies the importance of the resin sorption effect on the hydrodynamic of the resin flow inside the mold cavity and fibrous preform.

Keywords RTM · Theoretical · Experimental · Polymer composite

M. J. Nascimento Santos · A. G. Barbosa de Lima (✉) · I. R. Oliveira
Department of Mechanical Engineering, Federal University of Campina Grande,
Av. Aprígio Veloso, 882, Bodocongó, Campina Grande, PB 58429-900, Brazil
e-mail: antonio.gilson@ufcg.edu.br

M. J. Nascimento Santos
e-mail: marianajulie@outlook.com

I. R. Oliveira
e-mail: rodrigues.iran@hotmail.com

J. M. P. Q. Delgado
CONSTRUCT-LFC, Faculty of Engineering (FEUP), University of Porto,
Porto, Portugal
e-mail: jdelgado@fe.up.pt

© Springer International Publishing AG 2018   251
J. M. P. Q. Delgado and A. G. Barbosa de Lima (eds.), *Transport Phenomena in Multiphase Systems*, Advanced Structured Materials 93,
https://doi.org/10.1007/978-3-319-91062-8_8

1 Introduction

1.1 Basic Theory of Porous Media

Heat and mass transfer and fluid flow through porous media and porous materials have been the subject of great interest to scientists and engineers for many years, because of its numerous technological applications in different areas such as, petroleum, agricultural, materials, mechanical, chemical, civil, biomedical and environmental engineering, among others [1–6].

Because of the great importance of studying porous media, there are several works devoted to this theme [6–14].

However, what is a porous medium? It is a medium that is formed by a solid phase (with regular and random structure) and one or more fluids phase (multiphase system). The solid and fluid phases can be continuous or dispersed [15]. The porous structure of the solid phase is filled by the fluid phase.

When two or more immiscible fluids (multiphase fluid system) exist into the pore space (which fills out the voids completely), they are separated by interfaces. In this interface, interfacial forces act producing discontinuities in density and pressure [16]. Besides, fluid motion is characterized by different velocities, thus interaction effects arise, which influence the response of the porous media, also due to different material properties of the constituents [17]. From the macroscopic point of view this can be classified of how a fluid-saturated porous medium. The fluid saturated conditions assume that any vacant space exists in the porous medium [18].

What are the basic differences between porous medium and porous materials? To help us better understand this question we notice that materials with porous structure are called porous media [1]. Then, in some cases, the basic difference between them is related to the pores size. For example, fruits, vegetables and grains are porous materials which contain small pores. However, a packed-bed of these particles is a porous media which contains large pores. Figure 1 illustrates the basic contents.

When a fluid flows through a porous media (in isothermal or non-isothermal conditions), it suffers two basic phenomena: convection and diffusion (heat and/or

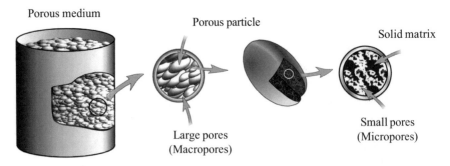

Fig. 1 Schematic of a porous medium and porous material

mass transfer). The simplest case, pure diffusion, occurs when the fluid is stagnant (no motion). Diffusion represents the spreading or mixing of the particles in the molecular level, where the motion is random (or Brownian in fluids) [19]. Convection is the motion of the particles in a macroscopic scale. It provokes a well-known phenomenon in the particles so called, dispersion, i.e., dispersion is associated with fluid flow.

According to Liu and Masliyah [19], the phenomenon of dispersion occurs due to the fluctuation of bulk flow, due to geometrical obstruction into the porous matrix, whereas the diffusion phenomenon is provoked by random molecular motion. The obstructions suffered by the fluid in the flow through a porous medium are caused by random distribution and dimension variations of the interconnected void space (capillary pathways). The magnitude of this effect in the flow dynamic is given by different physical parameters such as: porosity, tortuosity and permeability of the porous matrix, and fluid viscosity.

Permeability is a measure of how easily a single phase fluid moves in a porous medium under the influence of a dynamic pressure gradient (absolute pressure gradient adjusted for gravitational effects) [20].

Effective porosity is the local fraction of the porous medium, which consists of inter connected pore space (voids) that is available to fluid motion relative to the solid skeleton. Isolated pore space, assumed relatively small, is not taken into account [17, 20]. A more general definition is reported by Nield and Bejan [21] and Bear [22]. According to these authors porosity can be defined as the fraction of the total volume of the porous media that is occupied by pores. Then, both pores connected and disconnected are considered.

Tortuosity can be defined as the ratio between the straight line distance (macroscopic distance) and the true path distance length between two points traveled by a fluid into porous medium (microscopic distance) [19, 23].

1.2   What Are Composite Materials?

A composite material can be defined as a multiphase material composed of one or more discontinuous phases (reinforcement) embedded in a continuous phase, so called, matrix [24, 25].

Advanced polymer composites are made of high-strength fibers and moderately high-temperature resins at very high fiber volume fractions [26]. In general, composites use thermoset resins, with particular reference to epoxies and polyesters. In relation to fibers, they can be synthetic (glass, carbon, kevlar, boron, etc.) and natural (vegetable, animal, etc.). Fiber structure and adhesion between fiber-resin (reinforcement and matrix) are the main responsible parameters by the mechanical properties of the composite materials.

Historically, composite systems have been an important source of study in the world, because of the wide applications in different areas, specially, polymer composite reinforced by fiber.

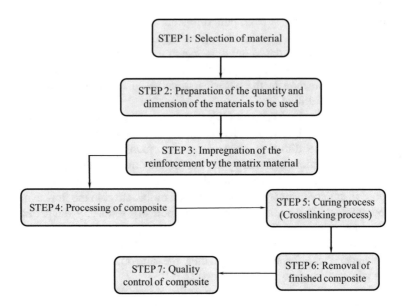

Fig. 2 Main steps for composite manufacturing

Based on the earlier information, this type of material can be classified as porous material or porous medium, depending upon the fiber content. There are many techniques to manufacture polymer composites, depending upon the type of applications, number of parts to be made, the geometry of the parts and the performance to be reached [27]. As an example, we can cite the following manufacturing techniques: Vacuum bag, hand lay-up, spray-up, filament winding, pultrusion and resin transfer molding (liquid injection family).

In general, the manufacturing processes cited above, have several basic steps in common. Obeying the following flow charts (Fig. 2).

The applications for advanced composites includes the sports materials, aerospace, aircraft, marine, biomaterials, ground transport, building and civil industries, in different segments such as: military jets (fuselage and wings), sports (tennis and squash rackets, skis, hockey sticks), marine (boats and sculls), ground-transport (vehicles, monorails, tramways and trains), façades, pipes, wind turbines, and many others [28–31].

2 Transport Phenomena Theory in RTM Processes

2.1 The Physics in RTM Processes

Resin Transfer Molding (RTM) Process is a type of fiber-reinforced polymer composite manufacturing process, in which the fiber preforms are placed in a closed cavity (mold) and the liquid polymer (resin) is impregnated to saturate the empty spaces (void pores) between the fiber of the preforms to make the composite structure [27]. The RTM technique can be explained by different steps, as illustrated in Fig. 3.

The phenomenon of resin injection into the mold cavity, in which the fibrous preform exists, is similar to the physical process of fluid flow in porous media (non-isothermal and reactive flow). In this step, parameters such as injection pressure, resin viscosity, porosity and permeability of the preform, temperature of the mold and resin, gates location on inlet resin and outlet air and resin, play important role in the control and optimization of the RTM process.

The resin reaction can be initiated early by heating (heat from the mold wall) or by mixing in a reacting compound prior to injection (catalyst material) [26].

In general, low injection pressure is applied, in order to avoid disturbing the placement of the fibers [28] and deformation in the mold (quality control of the parts).

As applying the RTM technique, fiber volume fraction can be in the range of 0.20 to 0.55; resin viscosity is in the range of 0.1 to 1 Pa/s; usual mold closing pressures range of 100–600 kPa [28].

After the injection step, is initiated the curing process of the resin. The time for the complete cure of the resin is around 6 to 30 min, depending on the cure kinetics of the mixture [32].

In the impregnation phase for manufacturing of polymer composite, the major problems that affect production quality and reproducibility (damage due to voids

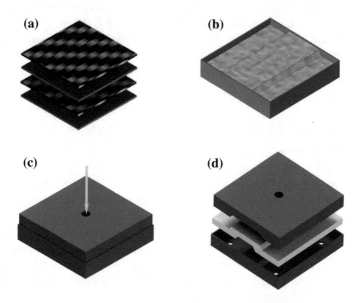

Fig. 3 Basic steps in RTM processes. **a** Preform, **b** preform in the mold, **c** closed mold and resin injection, and **d** demolding and final processing

and dry spots) are attributed to defects induced by the bad resin flow. The formation of micro-voids among fibers and dry spots are potential starting points for poor adhesion between fiber and resin, and the propagation of cracks and delamination, which reduce quality of the product post-processing.

Resin flowing through the porous medium (perform) suffers flow resistance between the fiber bundles and inside the fiber bundles (which can be measured in terms of two different permeabilities). Thus, we notice that the resin flow occurs in macro- and micro-scales. According to Laurenzi and Marchetti [33], in the macro-scale, macro-voids can be formed for different physical situations: (a) when the air displaced by the resin flow remains trapped, (b) when injection pressure is lower to overcome the resistance of the perform, and (c) when a resin with high viscosity is used. In micro-scale, the porosity is given by micro-voids. Thus, is necessary the total air evacuation at the beginning of the filling process, in order, to avoid it remains trapped in the tow producing micro-voids at the center of the fiber bundles. Further, micro-voids formation can be also due to the macroscopic pressure drop.

Since that the fiber bundles act as fluid sinks, so, fluid velocity is reduced thus, altering local pressure. Then, the pressure drop produces an apparent change in the permeability along the preform, which can be explained by the resin sink effect.

All discussion can be explained by the concept of the dual scale porous media. On the microscopic analysis, the individual fibers of the bundle are separated by a distance much smaller than that existing between two bundles. Thus, the resin flows more easily between the fiber bundles rather than inside one of them. Then, it is expected that permeability in the macropores are higher than the micropores. Consequently, due to different flow behavior, the resin continues to impregnate the bundle even when it has passed around it. This means that a part of the injected resin penetrates into the fibers, rather than push forward the resin flow front. The result is that the local pressure is affected by the resin flow rate inside a bundle when compared to that resin flow rate between bundles [33].

2.2 Heat Transfer and Fluid Flow Mathematical Formulation

The dominant transport phenomena during the application of the RTM technique for composite manufacturing are heat transfer and fluid flow due to the cure process and impregnation of the fibrous preform by the resin.

Because of the high costs involved in the RTM process, numerical simulations appear as an excellent alternative to be used for control and optimization of the filling process, and to reduce costs of the process. Then, we need an appropriated mathematical modeling to predict the filling process of the mold. This topic will be discussed in the next sections.

2.2.1 Fluid Flow Model

Because of the similarity, the mold filling process by resin is modeled considering the fluid flow through porous media theory. For this, Darcy's law (momentum conservation equation) in conjugation with mass conservation equation have been used to predict the location of the resin flow front and the fluid pressure distribution as it flows through the fibrous preform.

Following, will be presented an advanced mathematical modeling related to fluid flow through porous media, with particular reference to RTM process, in which exists the fluid absorption phenomena by fibrous preform.

For modeling of the fluid flow through the fibrous preform including fluid sorption by the fiber, we begin with the one of the appropriated mass conservation equation as follows [16, 21, 34–38].

$$\frac{\partial}{\partial t}(\varepsilon\rho) + \nabla \cdot \left(\rho\overrightarrow{U}\right) = S^M \tag{1}$$

where ε represents the porosity; ρ and $\overrightarrow{U}$ represent the density and mean (superficial) velocity vector of the fluid, respectively, t is the time of the fluid, and S^M represents the source term related to mass flow rate of resin absorbed by the fiber per volume unit.

The simplified momentum equation as applied to fluid flow in porous media is so called generalized Darcy's law. It is given as follows:

$$\overrightarrow{U} = -\frac{K}{\mu}\nabla P \tag{2}$$

where K represents the permeability of the porous media, μ is the resin viscosity, and P′ is the modified fluid pressure.

The modified fluid pressure taken into account the contribution of the fluid pressure and gravity effects as follows:

$$P' = P + \rho g z \tag{3}$$

where P is the fluid pressure, g is the acceleration due to gravity, and z is the height above a reference point.

Because of the existence of connected void space into the preform, the true (physical) velocity of the fluid into the pores is more than the superficial (mean) velocity. The relationship between these parameters, at any location into the porous medium, is given as follows:

$$\overrightarrow{U} = \varepsilon\overrightarrow{V} \tag{4}$$

where $\overrightarrow{V}$ represents the true fluid velocity vector.

2.2.2 Heat Transfer Model

Heat transfer that occurs during fluid flow and curing process can be modeled with the energy conservation equation as follows:

$$\frac{\partial}{\partial t}\left(\rho c_p T\right) + \nabla \cdot \left(\rho c_p \overrightarrow{V} T\right) = \nabla \cdot (k\nabla T) + S^H \tag{5}$$

where Cp and k represent specific heat and thermal conductivity of the mixture, respectively. These thermal parameters can be given using the mixture rules as follows:

$$\rho = \varepsilon\rho_f + (1 - \varepsilon)\rho_s \tag{6}$$

$$c_p = \varepsilon c_{p_f} + (1 - \varepsilon)c_{p_s} \tag{7}$$

$$k = \varepsilon k_f + (1 - \varepsilon)k_s \tag{8}$$

where the subscript f and s represent the fluid (resin) and solid (preform), respectively.

The energy equation considers the existence of local thermal equilibrium. This approach states that the resin and fiber temperatures are locally equal, at any moment of the resin injection and curing process. Besides, in the curing process the velocity vector $\overrightarrow{V}$ is null. Then, we have heat transfer by convection neglected and pure heat conduction phenomenon occurs.

The last term of the energy equation represents the rate of heat generation by chemical reaction, during curing process of the resin. It can be modeled as follows:

$$S^H = H_r \frac{d\alpha}{dt} \tag{9}$$

where H_r is the total heat of reaction and α represents the degree of cure.

The degree of cure α of the resin is given as follows [32, 39].

$$\alpha = \frac{Q}{H_r} \tag{10}$$

where Q is the heat evolved from time t = 0 to time t; it is predicted by the kinetic model. In this sense, Lee et al. [40] report information about heat of reaction, degree of cure and viscosity of a specific resin.

3 Application of RTM for Composite Manufacturing

3.1 Experimental Investigation

Oliveira [41] and Oliveira et al. [42] conducted one rectilinear injection experiment of orthophthalic polyester resin in a glass fiber mat (450 g/m^2) inserted in a stainless steel mold using a RTM system (room temperature and maximum injection pressure 0.25 bar). The mold with cavity dimensions of 320 × 150 × 3.6 mm, has one inlet and two outlet points (vents), and the top is made of glass to enable viewing of the advance of fluid flow. The system has a camera positioned above the mold for monitoring the impregnation process with the aid of a timer. The resin's viscosity was measured in a Brookfield viscometer HBDV-II + C/P with the S51 spindle. The following infiltration process conditions were used: (a) resin density $\rho = 1190$ kg/m^3; (b) fiber volume fraction Vf = 24%, (c) porosity $\varepsilon = 0.76$; (d) porous media permeability $k = 3.37 \times 10^{-10}$ m^2, and (e) fluid viscosity $\mu = 330$ cP. The RTM equipment is shown in Figs. 4 and 5 shows the preform inserted in the mold. Details about the equipment and experimental procedure (Fig. 4) can be found in the references cited. Figure 6 shows the resin flow front advancing into the mold.

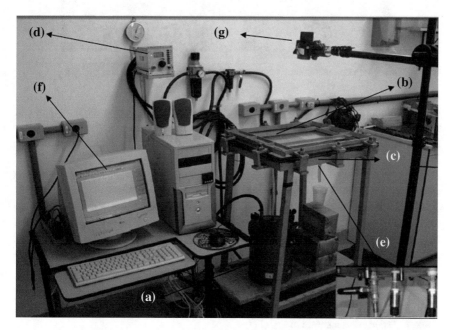

Fig. 4 Photo of the rectilinear RTM experimental apparatus from LACOMP/UFRGS (Brazil): **a** pressure vessel, **b** strengthened glass top mold, **c** steel bottom mold, **d** pressure controller, **e** pressure transducers, **f** data acquisition system and **g** camera

(a)   **(b)**

Fig. 5 **a** Stainless steel RTM mold without the preform and the top glass. **b** Preform inserted into the mold

Fig. 6 Flow front positions for the case pure resin with $P_{inj} = 0.25$ bar and t = 300 s

The injection pressure of the resin in the mold was monitored with pressure transducers during each experiment. The conditions of the experimental run, and the observed mold filling time and the calculated permeability data are presented in Table 1.

The extractions of new information from experimental data play an important role in different fields of engineering and science.

In this sense, parameter estimation techniques (inverse problem) arrive as an excellent mathematical tool which allows an efficient use of experimental data. The aim is to obtain process parameters that appear in a specific proposed mathematical modelling, and also to help in obtaining a better model that predict the behavior of a specific variable under investigation [16].

Table 1 Estimated parameters of Eq. (11)

a_1 (bar)	a_2 (–)	a_3 (bar)	a_4 (s⁻¹)	a_5 (–)	a_6 (s⁻¹)	a_7 (bar)
7.923739	−0.003113	0.003379	42.03634	50.54939	9.168526	−7.87010

From the collected injection pressure data and on the basis of the literature [43], a non-linear regression was made by Santos and Lima [44] using the Statistic® software, and the Rosenbrock and quasi-Newton method, yielding a pressure equation as a function of the process time as follows:

$$P_{inj}(t) = \begin{cases} a_1 t^{a_2} + a_3 \exp\left(\dfrac{a_4 t}{a_5 + a_6 t}\right) + a_7, & \text{for } 0 \leq t \leq t_e \\ P_e, & \text{for } t > t_e \end{cases} \tag{11}$$

where P_e is the final pressure (stable pressure) achieved in the experiment and t_e is the instant of time where P is now considered P_e.

3.2 Theoretical Procedure

3.2.1 The General Governing Equations

In problems of fluid flow through the fibrous preform including fluid sorption by the fiber, the mass conservation equation is given by Eq. (1). However, if the fluid is considered incompressible (constant density), Eq. (1) can be simplified and written as follows:

$$\nabla \cdot \vec{U} = -s \tag{12}$$

where $s = S/\rho$, corresponds to the sink term due to delayed saturation of fibrous preform compared to the empty spaces (pores) between the surrounding fibers.

Neglecting the gravitational effects, and substituting the Darcy's Law, Eq. (2) into Eq. (12), we obtain:

$$\nabla \cdot \left(-\frac{k}{\mu}\nabla P\right) = -s \tag{13}$$

By considering thermo-physical properties k and μ constants, the Eq. (13) takes the form:

$$\nabla^2 p = s\frac{\mu}{k} \tag{14}$$

To solve the Eq. (14), the following initial and boundary conditions can be used:

(a) $p = p_{inj}$ at the injection point;
(b) $\frac{\partial p}{\partial n} = 0$ on the walls (n is the normal direction to the wall), and
(c) $p = p_{ff}$ at the front line of the fluid. In general, p_{ff} is considered zero (gauge).

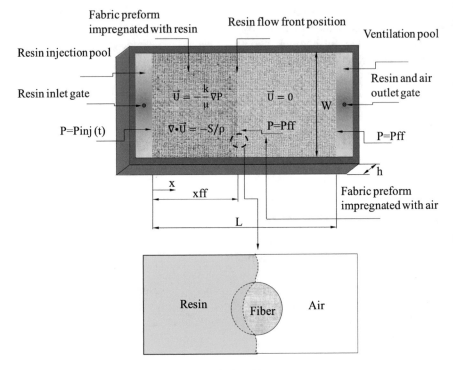

Fig. 7 Geometrical configuration of the physical problem

3.2.2 Physical Problem and the Geometry

Herein, we are giving emphasis for rectilinear infiltration problem. In this physical situation, the fluid is introduced through an inlet port at the border of the mold. The fluid flow is limited by the parallel wall, and exits the mold through the ventilation points at the other border. Figure 7 illustrates the physical problem discussed here.

3.2.3 Rectilinear Infiltration Model

For this simplified formulation, the following assumptions were considered:

(a) At the micro level, the porous medium is constituted of incompressible solid and fluid phases.
(b) The porous medium is fully saturated, i.e., the porous volume is completely filled with the fluid.
(c) The porous medium is considered homogeneous and isotropic.
(d) Thermo-physical properties are invariant with location and time.
(e) The fluid phases (air and resin) stay separated inside the porous media.

(f) Herein neither a geometrical interpretation of the pore structure nor the exact location of the individual components of the constituents, are considered.
(g) All the phases are in thermodynamic equilibrium.
(h) All the process is isothermal. This means that the energy balance equation is no more necessary.

For a one-dimensional flow of a Newtonian and incompressible fluid through porous media, the Darcy's law can be written as follows:

$$u_x = \frac{Q_x}{A} = -\frac{k}{\mu}\frac{dp}{dx} \tag{15}$$

where Q_x is the volumetric flow rate, A is the transversal section area of the mold cavity normal to the flow direction, u_x is the superficial velocity or the velocity based on an empty mold cavity, dP/dx is the pressure gradient of the fluid along the reinforcement and x represents the distance in the direction of the forward flow displacement.

In terms of interstitial velocity (u_x) or the flow front velocity (real velocity), we can write the superficial velocity as follows:

$$u_x = \varepsilon v_x = \varepsilon\frac{dx}{dt} \tag{16}$$

where ε is the porosity of the fibrous medium, and t is the time.

For a rectilinear and one-dimensional flow, which has fluid velocity equal to zero in the directions y and z, we can write Eq. (13) as follows:

$$\frac{d}{dx}\left(\frac{k}{\mu}\frac{dP}{dx}\right) = s \tag{17}$$

Then, considering the parameter s constant and integrating two times the Eq. 17, we obtained:

$$P = \frac{s\mu}{k}\frac{x^2}{2} + C_1 x + C_2 \tag{18}$$

where C_1 and C_2 are integration constants, that can be obtained by applying the following boundary conditions:

$$x = 0 \Rightarrow P = P_{inj}(t) \tag{19}$$

$$x = x_{ff} \Rightarrow P = P_{ff} = 0 \tag{20}$$

where P_{inj} is the injection pressure of the resin, x_{ff} and P_{ff} are the position and pressure of the front flow, respectively.

Then, substituting the Eqs. (19) and (20) into Eq. (18), is obtained:

$$C_2 = P_{inj}(t) \tag{21}$$

$$C_1 = -\frac{s\mu}{k}\frac{x_{ff}}{2} - \frac{P_{inj}(t)}{x_{ff}} \tag{22}$$

where P_{inj} is the injection pressure of the resin, x_{ff} and P_{ff} are the position and pressure of the front flow, respectively.

Replacing Eqs. (21) and (22) into Eq. (18), the Eq. (23) is obtained, which means that, if the viscosity of the fluid infiltrating an isotropic reinforcement (constant permeability) remains constant, there is a parabolic pressure distribution between the injection point and the flow front.

$$P = \frac{s\mu}{k}\frac{x^2}{2} - \left(\frac{s\mu}{k}\frac{x_{ff}}{2} + \frac{P_{inj}(t)}{x_{ff}}\right)x + P_{inj}(t) \tag{23}$$

Derivate of the Eq. (23) will be as follows:

$$\frac{dp}{dx} = \frac{s\mu}{k}x - \frac{s\mu}{k}\frac{x_{ff}}{2} - \frac{P_{inj}(t)}{x_{ff}} \tag{24}$$

For x = 0, we obtain the following pressure gradient:

$$\left.\frac{dP}{dx}\right|_{x=0} = -\frac{s\mu}{k}\frac{x_{ff}}{2} - \frac{P_{inj}(t)}{x_{ff}} \tag{25}$$

and for x = x_{ff}, we obtain:

$$\left.\frac{dP}{dx}\right|_{x=x_{ff}} = -sx_{ff} + \left(\frac{s}{2}x_{ff} + \frac{k}{\mu}\frac{P_{inj}(t)}{x_{ff}}\right) \tag{26}$$

Then, from the Eq. (15) and Fig. 6 we can write:

$$u_x(x = 0, t) = \frac{Q(t)}{Wh} = \frac{s}{2}x_{ff} + \frac{k}{\mu}\frac{P_{inj}(t)}{x_{ff}} \tag{27}$$

Then, injection volumetric flow rate will be given as follows:

$$Q_{inj}(t) = \frac{sWh}{2}x_{ff} + \frac{kWh}{\mu}\frac{P_{inj}(t)}{x_{ff}} \tag{28}$$

From Eq. (16), we can write:

$$u_x(x = x_{ff}) = \varepsilon \frac{dx_{ff}}{dt} = -sx_{ff} + \left(\frac{s}{2}x_{ff} + \frac{k}{\mu}\frac{P_{inj}(t)}{x_{ff}}\right) \tag{29}$$

Or yet,

$$x\frac{dx}{dt} = \left[\frac{-s}{2\varepsilon}\right]x^2 + \left[\frac{k}{\mu\varepsilon}P_{inj}(t)\right] \tag{30}$$

Then, solving the Eq. (30) we obtain the following results:

$$x_{ff}(t) = \sqrt{\frac{2k}{\mu\varepsilon}\left(e^{\frac{-s}{\varepsilon}t_{ff}}\int_0^{t_{ff}} e^{\frac{s}{\varepsilon}t}P_{inj}(t)dt\right)} \tag{31}$$

From Eq. (31), the time t_{ff} required for a fluid to attain a certain defined position x_{ff} in the mold is obtained, or vice versa. However, it is necessary to know pressure injection P_{inj} as a function of the filling time.

Substituting the Eq. (29) in the Eq. (16), we obtain the following equation for the interstitial velocity of the resin flow front:

$$V_x(x = x_{ff}) = \left(\frac{-s}{2\varepsilon}\right)x_{ff} + \frac{k}{\mu\varepsilon}\frac{P_{inj}(t)}{x_{ff}} \tag{32}$$

From Eq. (32), two possibilities of experimental conditions can be used to keep the rectilinear flow: the injection pressure or injection velocity (in terms of a volumetric flow rate). In this work only the first case will be used.

3.3 Results Analysis

Figure 8 shows the transient behavior of resin pressure at the injection gate compared to the experimental data reported in the literature [41, 42]. From an analysis of the figure it is noticed that an excellent fit was obtained with a correlation coefficient more than the 0.99 and a variance explained more than the 98% [44]. Table 1 summarize the parameters of Eq. (11) obtained with the non-linear regression.

It was verified that the pressure, during the first minutes of the injection process increases quickly tending to a steady state condition assuming a value almost constant of Pe = 21,801.8 Pa (te = 400 s). It happens due to the increase in the amount of the resin inside the mold with increasing injection time, which provokes an increase in the fluid flow resistance inside the porous media.

Figure 9 illustrates a comparison between simulated and experimental results of the resin front position inside mold cavityas function of the injection time. Analyzing the figure, it can be verified that, although the proposed mathematical

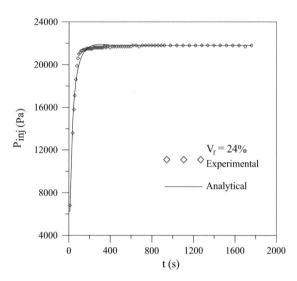

Fig. 8 Transient behavior of the resin pressure at the injection gate

formulation is simple: one-dimensional and transient formulation, an excellent agreement was obtained between the results, with a local maximum error at the front position of the resin of 10.72% considering a resin absorption term of $s = 0.0001 \text{ s}^{-1}$ and an injection time of 96 s. This error is a value lesser than the 11.7246% obtained without the consideration of sorption phenomena (Fig. 9). It was verified small deviations between the predicted and experimental data which was attributed to the fact that at practice, the flow occurs in the presence of a slightly difference in permeability and viscosity from the one at the beginning of the

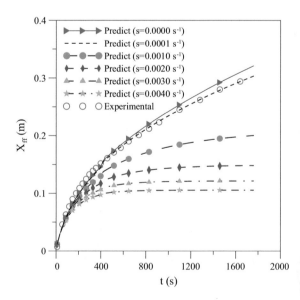

Fig. 9 Resin flow front position as a function of the injection time

Fig. 10 Square of the resin flow front position as a function of the injection time

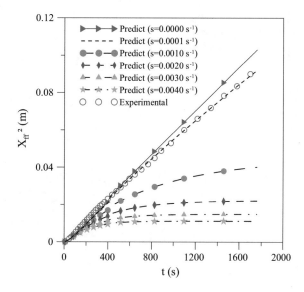

process. It is also noticed an approximately parabolic behavior of the resin flow front position as a function of the injection time, except in the case without to consider the sorption effect (Fig. 10). By analyzing the Fig. 9, at the end of the process (t = 1758 s), the local error was 0.2469% with the consideration of the sorption phenomena and 5.2778% without the consideration of this phenomena.

Figure 11 illustrates the pressure distribution inside the preform as a function of the resin flow front position at several instants of the injection process. As expected,

Fig. 11 Pressure distribution as a function of the resin front position at different process times

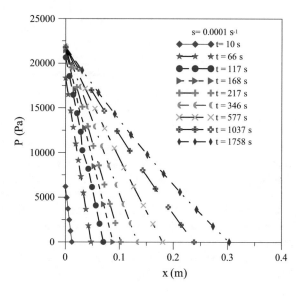

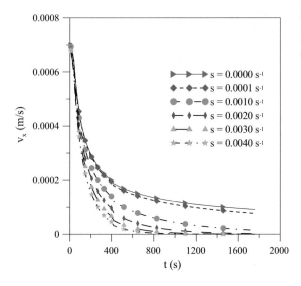

Fig. 12 Resin interstitial velocity behavior inside the mold as a function of the injection time

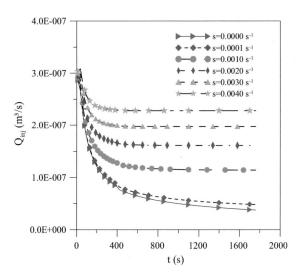

Fig. 13 Resin volumetric flow rate behavior at the inlet gate as a function of the injection time

it was verified a parabolic behavior of this parameter at any time of the process (Eq. 23). However, it can be observed, at the initial stages of the injection process, higher pressure gradients within the mold. It occurs because two effect: an increase in resin pressure at the injection gate with the time, which tends to stabilize for long times, as shown in Fig. 7, and the quantity of the resin inside the mold, that increases over time. We notice that when the sink term is null, pressure behavior is linear.

Figure 12 shows the real/interstitial velocity profile at the resin flow front position and Fig. 13 illustrates the injection volumetric flow rate at the inlet gate, both as a function of the injection time. From an analysis of these figure is noticed an asymptotic decrease with, both, time of injection, tending to a constant value for long times of process, depending on the sink term value. A more intense sorption effect (higher value of the parameter S) provokes more decay (high value of the velocity temporal variation) in the fluid velocity tending for an approximately null value in short injection time. Those instants correspond to the moments when the injection pressure becomes constant and the mold is completely filled by the resin.

4 Concluding Remarks

This chapter is devoted to fiber-reinforced composite manufacturing by liquid injection molding, with particular reference to resin transfer molding process (RTM). This class of physical problem is similar to heat transfer and fluid flow in porous media.

The interest in this type of problem is motivated by its importance in many practical situations related to different areas of science and engineering.

In this work, an appropriated review of porous media, composite materials and transport phenomena theory related to RTM process has been presented. From these information we can note that up to the present day, most of the theoretical studies of RTM process are based on numerous and severe approximations and assumptions in the modeling. One of them is: resin sorption effect negligible. Herein, advanced topics related to composite manufacturing by RTM process, including an advanced macroscopic multiphase mathematical modeling, which incorporate the effect of fluid sorption by fibers and the respective exact solution are presented and discussed.

Further, a mathematical analysis on the basis of the predicted results, and experiments of RTM process with rectilinear infiltration of resin into a mold cavity containing a fibrous preform is performed.

The advanced macroscopic fluid flow governing equations (with the effect of resin sorption) applied to porous media proved to be suitable to study RTM process with great success.

From the obtained results, the following conclusions may be derived:

(a) Predicted results have shown to be very accurate, with errors below 0.25% in the measurement of the resin flow front final position as compared to experimental data.
(b) The fluid flow front position and pressure change with an approximated parabolic behavior within the porous media, except for the case without sorption effect.
(c) Both flow front velocity and injection volumetric flow rate have presented an asymptotic behavior, tending to a constant value for long times.

Despite the good accuracy presented by the proposed model, there are some features that need comments:

(a) We have imposed that the solid skeleton is materially incompressible, thus neglecting the intrinsic compressibility of the solid material itself in comparison to the bulk compressibility of the porous skeleton. This assumption generally holds, except of the range of very small porosity values [45].

(b) We have assumed that the porous media is homogenous, however, depending on the reinforcement distribution, this is no true, and non-uniform distribution is verified. Thus, we have variable porosity inside the porous medium.

(c) We have ignored the effect of any heat transfer during the injection process. With respect to this assumption we start that during resin flow itself starts the curing process. Thus, we have an existing exothermic reaction, consequently heat transfer occurs and the fluid viscosity is modified.

Finally, it is interesting to point out that the findings obtained with this research may help researches and engineers in their studies about composite manufacturing especially that related to resin transfer technique.

Acknowledgements The authors thank to CNPq, FINEP and CAPES (Brazilian Research agencies) for financial support, and to the authors referred in this text that contributed for improvement of this work.

References

1. Hsu, C.-T.: Dynamic modeling of convective heat transfer in porous media. In: Vafai, K. (ed.) Handbook of Porous Media, 2nd edn, pp. 39–80. Taylor & Francis, Boca Raton, USA (2005)
2. Diebels, S.: Micropolar mixture models on the basis of the theory of porous media. In: Ehlers, W., Bluhm, J. (eds.) Porous Media: Theory, Experiments and Numerical Applications, pp. 121–145. Springer-Verlag, Heidelberg, Germany (2010)
3. Kowalski, S.J.: Mechanical aspect on drying of wet porous media. In: Ehlers, W., Bluhm, J. (eds.) Porous Media: Theory, Experiments and Numerical Applications, pp. 169–197. Springer-Verlag, Heidelberg, Germany (2010)
4. Larsson, R., Larsson, J., Runesson, K.: Theory and numerics of localization in a fluid-saturated elasto-plastic porous medium. In: Ehlers, W., Bluhm, J. (eds.) Porous Media: Theory, Experiments and Numerical Applications, pp. 315–340. Springer-Verlag, Heidelberg, Germany (2010)
5. Freij-Ayoub, R., Mühlhaus, H.-B., Probst, L.: Multicomponent Reactive Transport Modelling: Applications to One Body Genesis and Environmental Hazards. In: Ehlers, W., Bluhm, J. (eds.) Porous Media: Theory, Experiments and Numerical Applications, pp. 416–435. Springer-Verlag, Heidelberg, Germany (2010)
6. Khilar, K.C., Fogler, H.S.: Migrations of Fines in Porous Media. Kluwer Academic Publishers, Dordrecht, The Netherlands (1998)
7. Spanos, T.J.T.: The Thermophysics of Porous Media. Chapman & Hall/CRC, Boca Raton, USA (2002)
8. Viera, M.A.D., Sahay, P.N., Coronado, M., Tapia, A.O.: Mathematical and Numerical Modeling in Porous Media: Applications in Geosciences. CRC Press, Boca Raton, USA (2012)

9. Ingham, D.B., Pop, I.: Transport Phenomena in Porous Media. Pergamon, Oxford, UK (1998)
10. Ingham, D.B., Pop, I.: Transport Phenomena in Porous Media II. Pergamon, Amsterdam, The Netherlands (2002)
11. Ingham, D.B., Pop, I.: Transport Phenomena in Porous Media III. Elsevier Ltda, Oxford, UK (2005)
12. Vadász, P.: Emerging Topics in Heat and Mass Transfer in Porous Media: From Bioengineering and Microeletronics to Nanotechnology. Springer, New York, USA (2008)
13. Delgado, J.M.P.Q.: Industrial and Technological Applications of Transport in Porous Materials. Springer-Verlag, Berlin, Germany (2013)
14. Delgado, J.M.P.Q., Lima, A.G.B., Silva, M.V.: Numerical Analysis of Heat and Mass Transfer in Porous Media. Springer-Verlag, Berlin, Germany (2012)
15. Viskanta, R.: Combustion and heat transfer in inert porous media. In: Vafai, K. (ed.) Handbook of Porous Media, 2nd edn, pp. 607–644. Taylor & Francis, Boca Raton, USA (2005)
16. Harris, S.D., Ingham, D.B.: Parameter identification within a porous medium using genetic algorithims. In: Vafai, K. (ed.) Handbook of Porous Media, 2nd edn, pp. 687–742. Taylor & Francis, Boca Raton, USA (2005)
17. Lancellotta, R.: Coupling between the evolution of a deformable porous medium and the motion of fluids in the connected porosity. In: Ehlers, W., Bluhm, J. (eds.) Porous Media: Theory, Experiments and Numerical Applications, pp. 199–225. Springer-Verlag, Heidelberg, Germany (2010)
18. Ehlers, W.: Foundations of multiphasic and porous materials. In: Ehlers, W., Bluhm, J. (eds.) Porous Media: Theory, Experiments and Numerical Applications, pp. 3–86. Springer-Verlag, Heidelberg, Germany (2010)
19. Liu, S.L., Masliyah, J.H.: Dispersion in porous media. In: Vafai, K. (ed.) Handbook of Porous Media, 2nd edn, pp. 81–140. Taylor & Francis, Boca Raton, USA (2005)
20. McKibbin, R.: Modeling heat and mass transport processes in geothermal systems. In: Vafai, K. (ed.) Handbook of Porous Media, 2nd edn, pp. 545–571. Taylor & Francis, Boca Raton, USA (2005)
21. Nield, D., Bejan, A.: Convection in Porous Media, 3rd edn. Springer, New York, USA (2006)
22. Bear, J.: Dynamics of Fluid in Porous Media. Dover Publications Inc., New York, USA (1972)
23. Allen III, M.B., Behie, G.A., Trangestein, J.A.: Multiphase Flow in Porous Media. Springer-Verlag, Berlin, Germany (1988)
24. Agarwal, B., Broutmann, L.J., Chandrashekha, K.: Analysis and Performance of Fiber Composites. Wiley, New Jersey, USA (2006)
25. Carvalho, L.H., Canedo, E.L., Farias Neto, S.R., Lima, A.G.B., Silva, C.J.: Moisture transport process in vegetable fiber composites: theory and analysis for technological applications. In: Delgado, J.M.P.Q. (ed.) Industrial and Technologica Applications of Transport in Porous Materials, pp. 37–62. Springer-Verlag, Berlin, Germany (2013)
26. Gutowski, T.G.: A brief introduction to composite materials and manufacturing processes. In: Gutowski, T.G. (ed.) Advanced Composites Manufacturing, pp. 5–41. Wiley, New York, USA (1997)
27. Advani, S.G., Hsiao, K.T.: Transport phenomena in liquid composites molding processes and their roles in process control and optimization. In: Vafai, K. (ed.) Handbook of Porous Media, 2nd edn, pp. 573–606. Taylor & Francis, Boca Raton, USA (2005)
28. Bunsell, A.R., Renard, J.: Fundamentals of Fibre Reinforced Composite Materials. IOP Publishing, London, UK (2005)
29. Beukers, A., Bersee, H., Koussios, S.: Future aircraft structures: From metal to composite structures. In: Nicolais, L., Meo, M., Milella, E. (eds.) Composite Materials: A vision for the Future, pp. 1–50. Springer-Verlag, London, UK (2011)
30. Halpin, J.C.: Opportunities for polymeric-based composite applications for transport aircraft. In: Nicolais, L., Meo, M., Milella, E. (eds.) Composite Materials: A vision for the future, pp. 51–67. Springer-Verlag, London, UK (2011)

31. Shenoi, R.A., Dulieu-Barton, J.M., Quinn, S., Blake, J.I.R., Boyd, S.W.: Composite materials for marine applications: key challengers for the future. In: Nicolais, L., Meo, M., Milella, E. (eds.) Composite Materials: A vision for the Future, pp. 69–89. Springer-Verlag, London, UK (2011)

32. Mazumdar, S.K.: Composites Manufacturing: Materials, Product and Process Engineering. CRC Press, Boca Raton, USA (2002)

33. Laurenzi, S., Marchetti, M.: Advanced composite materials by resin transfer molding for aerospace applications, Chapter 10. In: Ning, H. (ed.) Composites and Their Properties, pp. 197–226. InTech, Rijeka, Croatia (2012)

34. McKibbin, R.: Mathematical models for heat and mass transport in geothermal systems. In: Ingham, D.B., Pop, I. (eds.) Transport Phenomena in Porous Media, pp. 131–154. Oxford, UK (1998)

35. Wang, C.Y.: Modeling multiphase flow and transport in porous media. In: Ingham, D.B., Pop, I. (eds.) Transport Phenomena in Porous Media, pp. 383–410. Oxford, UK (1998)

36. Bories, S., Prat, M.: Isothermal nucleation and bubble growth in porous media at low supersaturations. In: Inhgam, D.B., Pop, I. (eds.) Transport Phenomena in Porous Media II, pp. 276–315. Pergamon, Amsterdam, The Netherlands (2002)

37. Baytaş, A.C., Baytaş, A.F.: Entropy generation in porous media. In: Ingham, D.B., Pop, I. (eds.) Transport Phenomena in Porous Media III, pp. 201–226. Elsevier Ltda, Oxford, UK (2005)

38. Ma, L., Ingham, D.B., Pourkashanian, M.C.: Application of fluid flows through porous media in fuel cells. In: Ingham, D.B., Pop, I. (eds.) Transport Phenomena in Porous Media III, pp. 418–440. Elsevier Ltda, Oxford, UK (2005)

39. Kardos, J.L.: The processing science of reactive polymer composites. In: Gutowski, T.G. (ed.) Advanced Composites Manufacturing, pp. 43–80. Wiley, New York, USA (1997)

40. Lee, W.I., Loss, A.C., Springer, G.S.: Heat of reaction, degree of cure and viscosity of Hercules 3501-6 resin. J. Compos. Mater. 16(2), 510–520 (1982)

41. Oliveira, I.R.: Infiltration of loaded fluids in porous media via rtm process: theoretical and experimental analyses. Doctorate thesis, Process in Engineering, Federal University of Campina Grande, Campina Grande, Brazil (2014)

42. Oliveira, I.R., Amico, S.C., de Lima, A.G.B., de Lima, W.M.P.B.: Application of calcium carbonate in resin transfer molding process: an experimental investigation. Materialwiss. Werkstofftech. 46, 24–32 (2015)

43. Luz, F.F., Amico, S.C., Souza, J.A., Barbosa, E.S., Lima, A.G.B.: Resin transfer molding process: fundamentals, numerical computation and experiments. In: Delgado, J.M.P.Q., Barbosa de Lima, A.G., Vázquez da Silva, M. (Org.) Numerical Analysis of Heat and Mass Transfer in Porous Media. Series: Advanced Structured Materials, 1st ed., vol. 27, pp. 121–151. Springer-Verlag, Heidelberg, Germany (2012)

44. Santos, M.J.N., Barbosa de Lima, A.G.: Manufacturing fiber-reinforced polymer composite using rtm process: an analytical approach. Defect Diffus. Forum 380, 60–65 (2017)

45. Ehlers, W., Markert, B., Klar, O.: Biphasic description of viscoelastic foams by use of anextended Ogden-type formulation. In: Ehlers, W., Bluhm, J. (eds.) Porous Media: Theory, Experiments and Numerical Applications, pp. 275–294. Springer-Verlag, Heidelberg, Germany (2010)

Description of Osmotic Dehydration of Banana Slices Dipped in Solution of Water and Sucrose Followed by Complementary Drying Using Hot Air

A. F. da Silva Júnior, W. Pereira da Silva, V. S. de Oliveira Farias, C. M. D. P. da Silva e Silva and Antonio Gilson Barbosa de Lima

Abstract This chapter presents four mathematical approaches to describe processes of osmotic dehydration and complementary drying of agricultural products with cylindrical geometry. For this, four solutions for diffusion equation (in cylindrical coordinates) were proposed, two analytical solutions and two numerical solutions. All the formalism necessary to obtain these four solutions were presented. The most suitable models to describe the osmotic pretreatment and the complementary drying were determined using data of the osmotic dehydration of banana in solutions with 40 °Brix of sucrose and at temperature of 40 °C and data of the complementary drying at 40 °C. Programs developed in the FORTRAN language were used for optimization processes. Finally, the results obtained for the four models in the optimization processes were compared in order to obtain the best model.

A. F. da Silva Júnior · V. S. de Oliveira Farias
Physics and Mathematics Department, Federal University of Campina Grande,
Olho D'Água da Bica, S/N, Cuité, PB 58175-000, Brazil
e-mail: aluiziofsj.ces@gmail.com

V. S. de Oliveira Farias
e-mail: vera.solange6@gmail.com

W. Pereira da Silva (✉) · C. M. D. P. da Silva e Silva
Physics Department, Federal University of Campina Grande,
Av. Aprígio Veloso, 882, Bodocongó, Campina Grande, PB 58429-900, Brazil
e-mail: wiltonps@uol.com.br

C. M. D. P. da Silva e Silva
e-mail: cleidedps@gmail.com

A. G. Barbosa de Lima
Department of Mechanical Engineering, Federal University of Campina Grande,
Av. Aprígio Veloso, 882, Bodocongó, Campina Grande, PB 58429-900, Brazil
e-mail: antonio.gilson@ufcg.edu.br

© Springer International Publishing AG 2018   273
J. M. P. Q. Delgado and A. G. Barbosa de Lima (eds.), *Transport Phenomena in Multiphase Systems*, Advanced Structured Materials 93,
https://doi.org/10.1007/978-3-319-91062-8_9

Keywords Diffusion equation · Numerical solution · Analytical solution
Mass transfer · Finite-volume

1 Introduction

After harvest, changes that deteriorate agricultural products are influenced, in one way or another, by internal water concentration and mobility. Therefore, as it is well known in the literature, the shelf life of agricultural products can be increased with the water removal. This removal can be performed, for instance, using hot air [1–4]. However, this process is expensive due to the phase change of the water from liquid to vapor state, since the latent heat of vaporization for this substance is very high. In order to reduce the cost of the process, pretreatments are usually performed to partially remove water before complementary drying using, for example, hot air. One of these pretreatments is the osmotic dehydration, which is a simple and inexpensive method of partial water removal. According to Falcão Filho et al. [5], the osmotic dehydration process consists in dipping product slices into a hypertonic solution (water and one or more solutes). The difference in water chemical potential between the food and the osmotic medium is the driving force for dehydration. Thus, three basic mass flows occur during the process, namely: (1) water exits from the product slices to the hypertonic solution; (2) solute migrates from the hypertonic solution to the product slices; and, to a lesser extent, (3) solute leaves the product slices to the hypertonic solution.

Because the method requires no phase change for water during the process, osmotic dehydration has considerable industrial importance as far as energy efficiency is concerned [6]. According to several authors, including Aires et al. [7], this method of water removal is a complex phenomenon that depends upon various factors, such as the composition and concentration of the osmotic agent in which the product slices are dipped, temperature, immersion time, presence or absence of agitation in the osmotic medium, presence or absence of other pretreatment or treatment, ratio between sample and solution, as well as the nature, size and geometry of the product to be dehydrated. In addition, composition and concentration of the osmotic agent are significant due to their influences on osmotic pressure [8].

Silva et al. [9], citing Yadav and Singh [10], point out that the advantages of osmotic dehydration are as follows: (1) It is a low-temperature water removal process and hence minimum loss of color and flavor occur, (2) flavor retention is higher when sugar or sugar syrup is used as osmotic agent, (3) enzymatic oxidative browning is prevented as the fruit pieces are surrounded by sugar, thus making it possible to retain good color with little or no use of sulfur dioxide, (4) acid removal and sugar uptake by the fruit pieces give a sweeter product than conventionally dried product, (5) it partially removes water and thus reduces water removal load at the dryer, (6) energy consumption is much lower as no phase change is involved, (7) it increases solid density due to solid uptake and helps in getting better-quality product in freeze drying, (8) if salt is used as osmotic agent, higher moisture content is allowed at the end of drying as salt uptake influences water sorption behavior in the product, (9) the product's textural quality is better after reconstitution, (10) the product's storage life is greatly enhanced, (11) simple equipment is required for the process.

As osmotic dehydration is a method that partially removes water, in general the removal process is completed with conventional drying using, for instance, hot air. In order to describe osmotic dehydration and complementary drying, a model is usually used. Among the mathematical models found in the literature, empirical equations, Reaction Engineering Approach (REA) and diffusion models can be cited. According to Silva et al. [11], the main advantage of diffusion models is the possibility of predicting the distribution of water and/or solute within the slices at any time, and this allows to analyze the stresses that can damage the product.

For the description of the osmotic dehydration process using a diffusion model, the boundary condition depends on some factors, such as the medium agitation. If the medium is subjected to mechanical agitation, usually the external resistance is disregarded and the boundary condition of the first kind describes the process well [12]. For the medium with no mechanical agitation, generally, the external resistance exists and the appropriate boundary condition is of the third kind [7]. For complementary drying, particularly for fruit slices, the boundary condition is generally of the third kind [13].

As in the osmotic dehydration water removal is partial, shrinkage is often disregarded [11]. In this case, usually simple analytical solutions of the diffusion equation can be used to describe the process, in which the mass diffusivities (for water and solute) are considered with constant value. For this physical situation, programs like Prescribed Adsorption—Desorption (boundary condition of the first kind) and Convective Adsorption—Desorption (boundary condition of the third kind) are found in the literature to describe the process. However, for a rigorous description of the process, the shrinkage, as well as variable mass diffusivities must be considered [7, 12, 14]. For this case, a numerical solution of the diffusion equation is usually used.

In the last ten years, several works on osmotic dehydration of agricultural products have been found in the literature. Among these products, the following can be mentioned: pineapple slices [5, 15], carrot [16, 17], acerola [18], pumpkin [19–21], kiwi [21], pear [21], melon [22], apples [7, 23, 24]; guava [8, 11], coconut [9] and banana [14, 25, 26]. Due to the industrial importance of this method in the partial removal of water, the objective of this work is defined in the following.

The objective of this work is to describe the process of osmotic dehydration of banana slices, followed by complementary drying with hot air, using diffusion models for the geometry of a finite cylinder. For osmotic dehydration and complementary drying, the appropriate boundary condition for the diffusion model is identified.

2 Materials and Methods

2.1 Diffusion Equation

A finite cylinder of radius R and height L is shown in Fig. 1, highlighting these dimensions.

Since a symmetric diffusion in relation to the y-axis is assumed, a position within the cylinder can be given by the coordinates (r, y), defined through the system of axes r and y with origin at the center of the cylinder, as shown in Fig. 1.

To describe osmotic dehydration or drying for banana slices, the two-dimensional diffusion equation in cylindrical coordinates (r, y) with origin in the center of the cylinder can be written as follows:

$$\frac{\partial \Phi}{\partial t} = \frac{1}{r}\frac{\partial}{\partial r}\left(r\Gamma^{\Phi}\frac{\partial \Phi}{\partial r}\right) + \frac{\partial}{\partial y}\left(\Gamma^{\Phi}\frac{\partial \Phi}{\partial y}\right) \tag{1}$$

In Eq. (1), Φ is a generic variable that represents water quantity, W, and sucrose gain, S; Γ^{Φ} is the effective water diffusivity, D_W, and effective sucrose diffusivity, D_S, t is the time, r and y define a position within the cylindrical slices. For a drying process, Φ represents the moisture content, X, and Γ^{Φ} is the effective mass diffusivity, D.

The diffusion equation solution method to be used depends on the initial and boundary conditions of the phenomenon to be described, as well as on whether shrinkage can be disregarded or not. Thus, the boundary condition of the third kind is mentioned below.

Fig. 1 Finite cylinder with radius R and height L

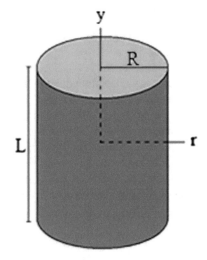

2.2   Boundary Condition of the Third Kind

Diffusion with convective boundary condition was one of the three models used to describe the drying process. The boundary condition of the third kind, also called Cauchy boundary condition, is defined by [27, 28]:

$$-\Gamma^{\Phi} \frac{\partial \Phi(r,y,t)}{\partial r}\Big|_{r=R} = h\left[\Phi(r,y,t)\big|_{r=R} - \Phi_{eq}\right] \tag{2}$$

and

$$-\Gamma^{\Phi} \frac{\partial \Phi(r,y,t)}{\partial y}\Big|_{y=\pm L/2} = h\left[\Phi(r,y,t)\big|_{y=\pm L/2} - \Phi_{eq}\right] \tag{3}$$

where h is the convective mass transfer coefficient and Φ_{eq} is the equilibrium value of Φ. Equation (2) refers to an infinite cylinder with radius R while Eq. (3) refers to an infinite slab with thickness L. The composition of these two simple geometries generates the finite cylinder, as shown in Fig. 2.

In Eqs. (2) and (3), the same value h was imposed, which means same resistance to the water flux in every surfaces.

Fig. 2 Intersection of infinite slab and infinite cylinder generating the finite cylinder

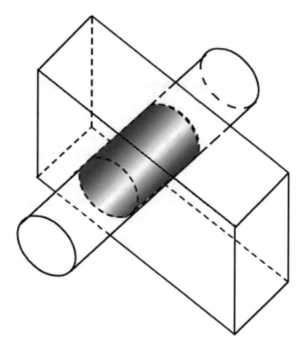

2.2.1 Analytical Solution for Boundary Condition of the Third Kind

For the boundary conditions given by Eqs. (2) and (3), the diffusion equation for the finite cylinder, given by Eq. (1), can be analytically solved to describe osmotic dehydration and drying processes, since the following assumptions can be accepted: (1) the dimensions of the finite cylinder do not vary during drying; (2) the initial distribution of Φ must be uniform; (3) diffusion is the only transport mechanism inside the cylinder; (4) the solid is considered as homogeneous and isotropic; (5) the effective mass diffusivity does not vary during drying; (6) the convective mass transfer coefficient is constant during drying; (7) osmotic dehydration or drying is considered under isothermal conditions.

For the assumptions above, the analytical solution of Eq. (1) in the position (r, y) at given instant t is [27, 28]:

$$\Phi(r, y, t) = \Phi_{eq} + (\Phi_0 - \Phi_{eq}) \sum_{n=1}^{\infty} \sum_{m=1}^{\infty} A_{n,1} A_{m,2} J_0 \left(\mu_{n,1} \frac{r}{R} \right) \cos \left(\mu_{m,2} \frac{y}{L/2} \right)$$
$$\times \exp \left[- \left(\frac{\mu_{n,1}^2}{R^2} + \frac{\mu_{m,2}^2}{(L/2)^2} \right) \Gamma^{\Phi} t \right] \tag{4}$$

where Φ_0 is the initial value of Φ. In Eq. (4), $A_{n,1}$ and $A_{m,2}$ are defined as:

$$A_{n,1} = \frac{2Bi_1}{J_0(\mu_{n,1}) \left(Bi_1^2 + \mu_{n,1}^2 \right)} \tag{5}$$

where J_0 is the Bessel function of first kind and zero order, and

$$A_{m,2} = (-1)^{m+1} \frac{2Bi_2 \left(Bi_2^2 + \mu_{m,2}^2 \right)^{1/2}}{\mu_{m,2} \left(Bi_2^2 + Bi_2 + \mu_{m,2}^2 \right)} \tag{6}$$

In Eqs. (5) and (6), Bi_1 and Bi_2 are the mass transfer Biot numbers for the infinite cylinder and infinite slab, respectively, as shown in Fig. 2. For a finite cylinder with radius R and height L, such Biot numbers are expressed as follows:

$$Bi_1 = \frac{hR}{\Gamma^{\Phi}}, \tag{7}$$

and

$$Bi_2 = \frac{h(L/2)}{\Gamma^{\Phi}}. \tag{8}$$

Since the same value h was imposed for all surfaces of the finite cylinder, Eqs. (7) and (8) result in:

$$Bi_2 = \frac{Bi_1\,(L/2)}{R}. \tag{9}$$

In Eqs. (4), (5) and (6), $\mu_{n,1}$ and $\mu_{m,2}$ are the roots of the characteristic equations for the infinite cylinder and infinite slab, respectively. These characteristic equations are given by:

$$\frac{J_0(\mu_{n,1})}{J_1(\mu_{n,1})} = \frac{\mu_{n,1}}{Bi_1}, \tag{10}$$

where J_1 is the Bessel function of first kind and first order, and

$$\cot \mu_{m,2} = \frac{\mu_{m,2}}{Bi_2}. \tag{11}$$

Equation (4) gives the value of Φ for any position (r, y) within the finite cylinder at a given instant t. The average value $\overline{\Phi}(t)$ at time t is given by:

$$\overline{\Phi}(t) = \Phi_{eq} + (\Phi_0 - \Phi_{eq}) \sum_{n=1}^{\infty} \sum_{m=1}^{\infty} B_{n,1} B_{m,2} \exp\left[-\left(\frac{\mu_{n,1}^2}{R^2} + \frac{\mu_{m,2}^2}{(L/2)^2}\right)\Gamma^\Phi t\right] \tag{12}$$

The coefficients $B_{n,1}$ and $B_{m,2}$ are given, respectively, by:

$$B_{n,1} = \frac{4Bi_1^2}{\mu_{n,1}^2\left(Bi_1^2 + \mu_{n,1}^2\right)}. \tag{13}$$

and

$$B_{m,2} = \frac{2Bi_2^2}{\mu_{m,2}^2\left(Bi_2^2 + Bi_2 + \mu_{m,2}^2\right)}. \tag{14}$$

For the boundary condition of the third kind, to determine the process parameters, Eq. (12) can be fitted to the experimental datasets available, through the optimization methodology proposed by Da Silva et al. [29], using the program called Convective Adsorption—Desorption Software.

2.2.2 Numerical Solution for Boundary Condition of the Third Kind

In the complementary drying process, the effects of shrinkage become non-negligible. Therefore, it is necessary to include in the mathematical modeling the effects of the shrinkage and, consequently, the variation of the effective diffusivity. For this, a numerical solution is required for Eq. (1). The methodology used to discretize the diffusion equation is the same as that used by Silva Junior et al. [13] and Silva Junior et al. [14].

In order to solve Eq. (1), the hypotheses (2), (3), (4), (6) and (7) (Sect. 2.2.1) were assumed. In addition to these hypotheses, it was assumed that the product dimensions and the effective mass diffusivity vary during the process.

The method used to discretize the diffusion equation was the Finite Volume Method using a fully implicit formulation [30]. Discretization begins by decomposing the continuous domain into a finite number of subdomains V_i, with i = 1,..., N, called control volumes. These control volumes are characterized by a point where the unknown variables are calculated, which are known as nodal points [31–33].

In Fig. 3, a continuous domain Ω is presented for the geometry of a finite cylinder, divided into 256 (16 × 16) control volumes. Among these control volumes, a volume of dimensions Δr and Δy is highlighted, which is characterized by the nodal point P.

In order to reduce the computational cost, only one quarter of the two-dimensional grid was considered, as Fig. 4 shows. For this, the spatial distribution of the quantity of interest inside the cylinder was assumed to have radial and axial symmetries with respect to the central axis.

The control volume set for each element of the grid is illustrated in Fig. 5.

By Fig. 5, it is possible to note that there are 9 distinct types of control volumes, which are characterized by the types of neighboring volumes and by the contact with the external medium. However, only the discretizations of an internal control volume (which is not in contact with external environment) and a control volume

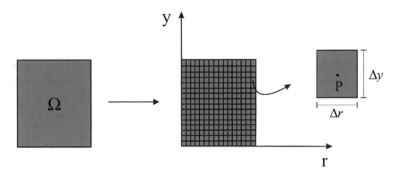

Fig. 3 Continuous domain divided into 256 (16 × 16) control volumes

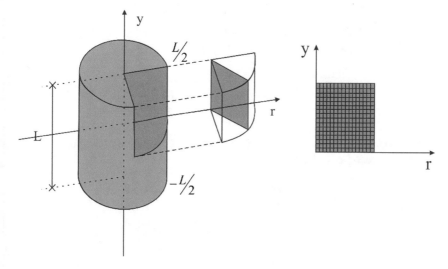

Fig. 4 Grid obtained with the hypothesis of symmetry

Fig. 5 Control volume for cylindrical geometry

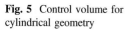

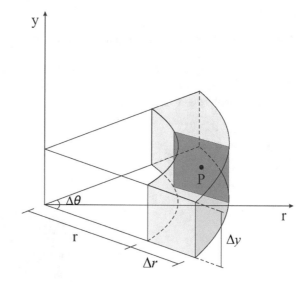

southeast (with zero flow to the south, in contact with the external medium to the east, and neighbors to the north and west) will be presented.

Equation (1) has been integrated into the space $r_P\Delta\theta\Delta r\Delta y$ and at time Δt, as shown in Eq. (15).

$$\int_{t}^{t+\Delta t} \int_{w}^{e} \int_{s}^{n} \frac{\partial \Phi}{\partial t} \Delta\theta r_P dy dr dt = \int_{t}^{t+\Delta t} \int_{w}^{e} \int_{s}^{n} \frac{1}{r} \frac{\partial}{\partial r} \left(r \Gamma^\Phi \frac{\partial \Phi}{\partial r} \right) \Delta\theta r_P dy dr dt$$

$$+ \int_{t}^{t+\Delta t} \int_{w}^{e} \int_{s}^{n} \frac{\partial}{\partial y} \left(\Gamma^\Phi \frac{\partial \Phi}{\partial y} \right) \Delta\theta r_P dy dr dt \tag{15}$$

By calculating the integrals and performing some algebraic manipulations, the following equation is obtained:

$$\left(\Phi_P - \Phi_P^0 \right) r_P \frac{\Delta r \Delta y}{\Delta t} = \left(r_e \Gamma_e^\Phi \frac{\partial \Phi}{\partial r} \bigg|_e - r_w \Gamma_w^\Phi \frac{\partial \Phi}{\partial r} \bigg|_w \right) \Delta y$$

$$+ \left(\Gamma_n^\Phi \frac{\partial \Phi}{\partial y} \bigg|_n - \Gamma_s^\Phi \frac{\partial \Phi}{\partial y} \bigg|_s \right) r_P \Delta r. \tag{16}$$

From Eq. (16), the discretized equations for all types of control volumes are obtained.

Discretization of the Diffusion Equation for the Internal Control Volumes

The internal control volumes have no contact with the external medium and have four neighboring control volumes, one to the north, one to the south, one to the west, and finally one to the east (Fig. 6).

Fig. 6 Internal control volume and control volumes to the north (N), to the south (S), to the west (W) and to the east (E)

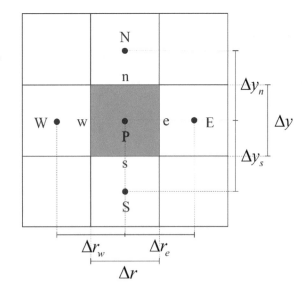

Considering a uniform grid, that is, $\Delta y_n = \Delta y_s = \Delta y$ and $\Delta r_w = \Delta r_e = \Delta r$, the following approximations of the derivatives are obtained:

$$\left.\frac{\partial \Phi}{\partial r}\right|_e \cong \frac{\Phi_E - \Phi_P}{\Delta r} \tag{17}$$

$$\left.\frac{\partial \Phi}{\partial r}\right|_w \cong \frac{\Phi_P - \Phi_W}{\Delta r} \tag{18}$$

$$\left.\frac{\partial \Phi}{\partial y}\right|_n \cong \frac{\Phi_N - \Phi_P}{\Delta y} \tag{19}$$

$$\left.\frac{\partial \Phi}{\partial y}\right|_s \cong \frac{\Phi_P - \Phi_S}{\Delta y} \tag{20}$$

Substituting Eqs. (17)–(20) into Eq. (16), the following equation is obtained:

$$\left(\Phi_P - \Phi_P^0\right) \frac{r_P \Delta r \Delta y}{\Delta t} = r_e \Gamma_e^\Phi \left(\frac{\Phi_E - \Phi_P}{\Delta r}\right) \Delta y - r_w \Gamma_w^\Phi \left(\frac{\Phi_P - \Phi_W}{\Delta r}\right) \Delta y$$
$$+ \Gamma_n^\Phi \left(\frac{\Phi_N - \Phi_P}{\Delta y}\right) r_P \Delta r - \Gamma_s^\Phi \left(\frac{\Phi_P - \Phi_S}{\Delta y}\right) r_P \Delta r$$

By grouping similar terms, the following algebraic equation is obtained:

$$A_P \Phi_P = A_E \Phi_E + A_W \Phi_W + A_N \Phi_N + A_S \Phi_S + B \tag{21}$$

where

$$A_P = r_P \frac{\Delta r \Delta y}{\Delta t} + r_e \Gamma_e^\Phi \frac{\Delta y}{\Delta r} + r_w \Gamma_w^\Phi \frac{\Delta y}{\Delta r} + r_P \Gamma_n^\Phi \frac{\Delta r}{\Delta y} + r_P \Gamma_s^\Phi \frac{\Delta r}{\Delta y} \tag{22}$$

$$A_e = r_e \Gamma_e^\Phi \frac{\Delta y}{\Delta r} \tag{23}$$

$$A_w = r_w \Gamma_w^\Phi \frac{\Delta y}{\Delta r} \tag{24}$$

$$A_n = r_P \Gamma_n^\Phi \frac{\Delta r}{\Delta y} \tag{25}$$

$$A_s = r_P \Gamma_s^\Phi \frac{\Delta r}{\Delta y} \tag{26}$$

$$B = r_P \frac{\Delta r \Delta y}{\Delta t} \Phi_P^0 \qquad (27)$$

Discretization of the Diffusion Equation for the Southeast Control Volume

The Southeast control volume has two neighboring volumes, one to the north and one to the west, one border in contact with the external medium and another on the axis of symmetry, as shown in Fig. 7.

As the boundary "s" is on the axis of symmetry, the following equality is considered:

$$\left.\frac{\partial \Phi}{\partial y}\right|_s \cong 0 \qquad (28)$$

Considering the equality of the diffusive and convective flows at the border "e", the following equation is obtained

$$\Phi_e = \frac{h_e \Phi_{\infty e} + \frac{2\Gamma_e^\Phi}{\Delta r} \Phi_P}{\frac{2\Gamma_e^\Phi}{\Delta r} + h_e} \qquad (29)$$

Substituting (29) into $\phi_e'' = h_e(\Phi_e - \Phi_{\infty e})$ one obtains:

$$\phi_e'' = \frac{(\Phi_P - \Phi_{\infty e})}{\frac{1}{h_e} + \frac{\Delta r}{2\Gamma_e^\Phi}} \qquad (30)$$

Fig. 7 Control volume to the southeast with its neighbors to the north and west

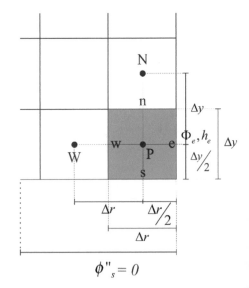

$$\phi''_s = 0$$

Through some algebraic manipulations, one obtains:

$$\Gamma_e^\Phi \frac{\Phi_e - \Phi_P}{\Delta r/2} = \frac{(\Phi_{\infty e} - \Phi_P)}{\frac{1}{h_e} + \frac{\Delta r_e}{2\Gamma_e^\Phi}} \tag{31}$$

Substituting (28), (31), (18) and (19) into (16), one obtains:

$$r_P \frac{\Delta r \Delta y}{\Delta t} \Phi_P - r_P \frac{\Delta r \Delta y}{\Delta t} \Phi_P^0 = r_e \frac{(\Phi_{\infty e} - \Phi_P)}{\frac{1}{h_e} + \frac{\Delta r_e}{2\Gamma_e^\Phi}} \Delta y - r_w \Gamma_w^\Phi \left(\frac{\Phi_P - \Phi_w}{\Delta r} \right) \Delta y$$

$$+ \Gamma_n^\Phi \left(\frac{\Phi_N - \Phi_P}{\Delta y} \right) r_P \Delta r.$$

By grouping similar terms, one obtains:

$$A_P \Phi_P = A_w \Phi_W + A_n \Phi_N + B, \tag{32}$$

where

$$A_P = r_P \frac{\Delta r \Delta y}{\Delta t} + \frac{r_e \Delta y}{\left(\frac{1}{h_e} + \frac{\Delta r}{2\Gamma_e^\Phi} \right)} + r_w \Gamma_w^\Phi \frac{\Delta y}{\Delta r} + r_P \Gamma_n^\Phi \frac{\Delta r}{\Delta y} \tag{33}$$

$$A_w = r_w \Gamma_w^\Phi \frac{\Delta y}{\Delta r} \tag{34}$$

$$A_n = r_P \Gamma_n^\Phi \frac{\Delta r}{\Delta y} \tag{35}$$

$$B = r_P \frac{\Delta r \Delta y}{\Delta t} \Phi_P^0 + \frac{r_e \Delta y}{\frac{1}{h_e} + \frac{\Delta r}{2\Gamma_e^\Phi}} \Phi_{\infty e} \tag{36}$$

2.2.3 Average Value of the Moisture Content

At each instant, the numerical solution gives the value of the moisture content at each nodal point. As the contributions of each control volume to the average value are not the same, the average value of moisture content, at every time instant, is obtained by calculating the weighted average:

$$\overline{X} = \frac{\sum X_P V_P}{\sum V_P} \tag{37}$$

in which $V_p = r_p \Delta \theta \Delta r \Delta y$.

2.2.4 Evaluation of Water Diffusivity

The diffusion equation was discretized assuming that the effective water diffusivity can vary with the local value of the moisture content, i.e.,

$$D = f(X, a, b),$$ (38)

where a and b are coefficients of a function that fit the numerical solution to the experimental data, and are determined by optimization.

Considering the discretization presented, one needs to know the parameter D at the interfaces of all control volumes. In the case of value D at the nodal points, this is calculated by means of Eq. (14), at every step of the optimization process. To calculate D at the interface between the control volumes P and E, for instance, the following expression is used for a uniform grid [30].

$$D_{eq} = \frac{2D_P D_E}{D_P + D_E}$$ (39)

2.3 Boundary Condition of the First Kind

2.3.1 Analytical Solution for Boundary Condition of the First Kind

If the external resistance to the mass flux is neglected, the Biot numbers Bi_1 and Bi_2 are considered infinite. In this case, Eq. (10) is written as:

$$J_0(\mu_{n,1}) = 0,$$ (40)

and Eq. (11) is given by:

$$\cot \mu_{m,2} = 0.$$ (41)

Thus, the solution $\overline{\Phi}$ (t) for the diffusion equation is still given by Eq. (12). However, the coefficients $B_{n,1}$ and $B_{m,2}$ are now given, respectively, by:

$$B_{n,1} = 4/\mu_{n,1}^2$$ (42)

and

$$B_{m,2} = 2/\mu_{m,2}^2$$ (43)

For the boundary condition of the first kind, to determine the process parameters, Eq. (12) can be also fitted to the experimental datasets available, through the optimization methodology proposed by Da Silva et al. [34], using the program called Prescribed Adsorption—Desorption Software.

2.3.2 Numerical Solution for Boundary Condition of the First Kind

The difference between the discretizations presented in Sect. 2.2.2 and the discretizations assuming a boundary condition of the first kind occurs in the equations related to the volumes in contact with the external medium. Therefore, the equation related to the internal control volumes is Eq. (21), presented in Sect. 2.2.2.

As already noted in the previous section, in a boundary condition of the first kind the surface resistance to mass flow is neglected. In this case the convective mass transfer coefficient is considered as tending to infinity. Thus, in Eqs. (33) and (36), one obtains:

$$A_P = r_P \frac{\Delta r \Delta y}{\Delta t} + \frac{2r_e \Delta y}{\Delta r} \Gamma_e^\Phi + r_w \Gamma_w^\Phi \frac{\Delta y}{\Delta r} + r_p \Gamma_n^\Phi \frac{\Delta r}{\Delta y} \tag{44}$$

$$B = r_P \frac{\Delta r \Delta y}{\Delta t} \Phi_P^0 + \frac{2r_e \Delta y}{\Delta r} \Gamma_e^\Phi \Phi_{\infty e} \tag{45}$$

Analogously, it is considered that $h_w = h_n = h_s$ also tend to infinity in the discretizations for the other seven types of control volumes. Thus, the other equations for the boundary condition of the first kind are obtained.

2.4 Optimizations

In order to obtain the process parameters (a, b and h) using experimental datasets, the optimizers existing in the programs Convective (analytical solution, boundary condition of the third kind), Prescribed (analytical solution, boundary condition of the first kind) and LS optimizer (any differential equation or function) were used [35–37]. These optimizers were developed from an inverse method in which initial values are assigned to the parameters and then these are corrected in order to minimize an objective function. The objective function used in this work was the chi-square, which is defined as follows:

$$\chi^2 = \sum_{i=1}^{N_p} \left[\overline{\Phi}_i^{exp} - \overline{\Phi}_i^{sim}(\Gamma^\Phi, h) \right]^2 \frac{1}{\sigma_i^2}, \tag{46}$$

where $\overline{\Phi}_i^{exp}$ is the average value of the quantity of interest relative to the ith experimental point, $\overline{\Phi}_i^{exp}(\Gamma^\Phi, h)$ is the average value of the quantity of interest calculated by the numerical solution as a function of Γ^Φ and h, N_P is the number of experimental points, and $\frac{1}{\sigma_i^2}$ is the statistical weight for the ith experimental point. Thus, the objective function depends on Γ^Φ and the convective mass transfer coefficient. If Γ^Φ is considered as variable, that is, $\Gamma^\Phi = f(\Phi, a, b)$, the chi-square will depend on the parameters a, b and h. In the case of the boundary condition of the first kind, $\overline{\Phi}_i^{sim}$ will depend only on Γ^Φ.

2.5 Osmotic Dehydration Experiments

The banana used in the osmotic dehydration experiments was the Manzano banana. This fruit belongs to the AAB genomic group, purchased in the local market (Campina Grande PB, Brazil) at the fourth stage of maturity: yellower than greener. After purchased, the bananas were kept at room temperature until they reached the last stage of maturity: yellow with brown spots. The initial moisture content of the samples after maturation was 3.320 (d.b.).

The methodology that will be described was proposed by Silva Junior et al. [14]. Osmotic dehydration experiments were carried out in the Laboratory of Storage and Processing of Agricultural Products of the Center for Technology and Natural Resources of the Federal University of Campina Grande. Before the osmotic dehydration experiments, the fruits were washed in running water and sanitized with chlorinated water for a period of 15 min. After sanitized, the fruits were washed again with running water and manually peeled. Finally, they were cut into slices approximately L = 1.0 cm thick (Fig. 8). The average radius was R = 1.7 cm.

After being cut, the samples were separated, weighed and placed in baskets, which were grouped in triplicate and labeled as 1.1, 1.2 and 1.3 to 11.1, 11.2 and

Fig. 8 Banana slice with 1.0 cm thickness

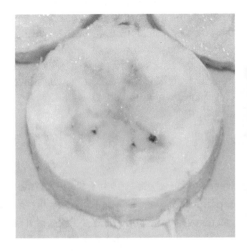

Fig. 9 Triplicate samples in baskets prepared for each instant of the osmotic dehydration process

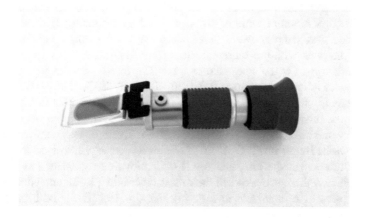

Fig. 10 Refractometer used to adjust the concentration of the solution

11.3. These labels represent the times in which the groups would be withdrawn from the solution (Fig. 9).

The solution used for osmotic dehydration was binary, distilled water and commercial crystal sugar, prepared with ratio of 1:15 (g/g) (fruit to solution) at concentration of 40 °Brix. This ratio had the objective of keeping the concentrations unchanged during the experiments. These concentrations were controlled through a refractometer (INSTRUTHERM, model RT-280), which is shown in Fig. 10.

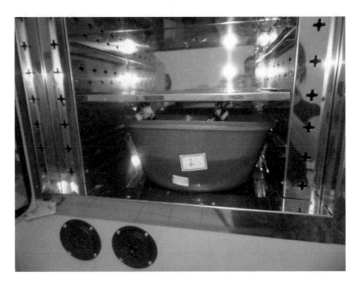

Fig. 11 Samples in solution being placed in the oven

Osmotic dehydration experiment was performed for a solution with concentration of 40 °Brix at temperature of 40 °C. Once the temperature of the experiment was reached, the samples were placed simultaneously in the solution and then placed in the oven with pre-set temperature (Fig. 11).

The variation of water and sucrose quantities in the product was monitored through each triplicate at the times 0, 10, 30, 60, 90, 120, 150, 180, 210, 240, 1440 and 1800 min, which were denoted by n_0, n_1,...,n_{11}. The sample in triplicate for time n_0 is the fresh sample, which was taken directly to the drying oven to determine its dry mass. At each time interval, the triplicate sample was removed from the solution, washed with distilled water (to remove sucrose adhered on its surface), and then lightly wiped with paper towel (to remove excess water), as shown in Fig. 12. Finally, these samples had their mass determined by an analytical balance and then placed (in crucibles) in a drying oven at 105 °C for 24 h in order to determine the dry masses.

Sucrose gain and water loss kinetic processes were described by the sample of the time n_{11}, which was withdrawn from solution at time 1800 min. To determine the total mass at all times, the following formula was used, obtained through a simple rule of three:

$$m_{11}^t = m_{11}^0 \frac{m_x^t}{m_x^0},\tag{47}$$

where m_{11}^t is the mass of the sample n_{11} at time t, m_{11}^0 is the mass of the sample n_{11} at time 0, m_x^t is the mass of the sample n_x at time t, and m_x^0 is the mass of the sample n_x at time 0.

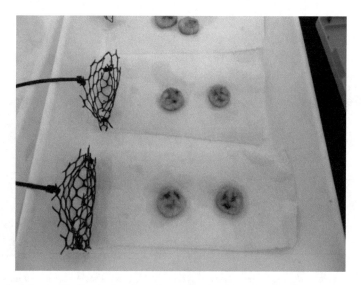

Fig. 12 Samples (and baskets) taken from the solution for determination of water loss and sucrose gain

To determine the dry mass at the other times, the following formula was used, also obtained through a simple rule of three:

$$ms_{11}^t = ms_x^t \frac{m_{11}^t}{m_x^t}, \tag{48}$$

where ms_{11}^t is the dry mass of sample n_{11} at time t, ms_x^t is the dry mass of sample n_x at time t, m_{11}^t the mass of sample n_{11} at time t, and m_x^t m_x^t is the mass of the sample n_x at time t.

To calculate the water quantity in percentage in the product at each time t, the following formula was used:

$$\bar{c}_w^t = \frac{m_{11}^t - ms_{11}^t}{m_{11}^0 - ms_{11}^0} \times 100, \tag{49}$$

where $\bar{c}_w^t$ is the percentage of water quantity in the product at time t, m_{11}^t is the mass of the sample n_{11} at time t, m_{11}^0 is the mass of the sample n_{11} at time zero, ms_{11}^t is the dry mass of the sample n_{11} at time t and ms_{11}^0 is the dry mass of the sample n_{11} at time zero. To calculate the sucrose quantity incorporated (in percentage) at time t over the initial dry mass, the following formula was used:

$$\bar{c}_s^t = \frac{ms_{11}^t - ms_{11}^0}{ms_{11}^0} \times 100, \tag{50}$$

(a)

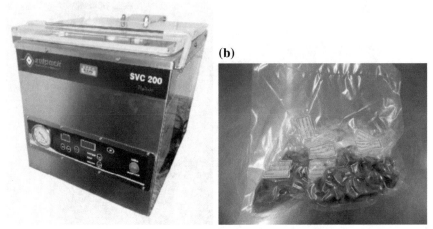

Fig. 13 **a** Device used for vacuum storage. **b** Samples of banana stored under vacuum

where $\vec{c}_s^t$ is the percentage of sucrose quantity in the product at time t.

After the osmotic dehydration process, the banana samples were stored under vacuum and then placed in a plastic box at room temperature in order to be submitted to the subsequent convective drying. The stored samples and the SVC 200 SULPACK vacuum sealer used for storage are shown in Fig. 13.

2.6 Complementary Drying

The methodology that will be described was proposed by Silva Junior et al. [13]. After the osmotic dehydration process, the samples were subjected to a complementary drying at temperature of 40 °C. Initially the baskets used for drying had their weights determined, and then seven samples were placed in each basket, as shown in Fig. 14.

Thereafter the weights of the baskets together with the samples were determined. Finally, the baskets with the samples were placed simultaneously in an oven with air circulation and renewal (AMERICANLAB, model AL 102/480) (Fig. 15), to start drying.

At intervals of 5, 10, 20, 30, 60, 120, 450 and 720 min samples in triplicate were removed from the oven and their weights were determined on an analytical balance. Also at these intervals, other samples in triplicate, used to track the shrinkage phenomenon, were removed from the oven and had their length (L) and radius (R) measured with a caliper. This procedure was repeated until constant weight was achieved. The initial values of the thickness and radius of the samples were

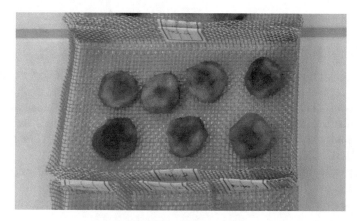

Fig. 14 Basket with osmotically dehydrated samples at 40 °C and 40 °Brix during drying at 40 °C

Fig. 15 Oven with air circulation and renewal used for convective drying

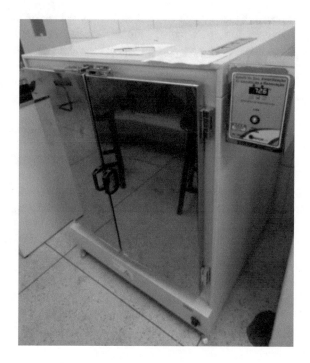

$L_0 = 0.67$ cm and $R_0 = 1.50$ cm (on average). After drying, the samples were placed in a drying oven at 105 °C for 24 h to determine the dry mass.

The experimental dataset obtained for convective drying kinetics was considered in dimensionless form using the following equation:

$$\bar{X}^*(t) = \frac{\bar{X}(t) - X_{eq}}{X_i - X_{eq}} \times 100, \tag{51}$$

where X_{eq} (d.b.) is the equilibrium moisture content, $\bar{X}$ (d.b.) is the average moisture content and X_i (d.b.) is the initial moisture content of the drying process, which is the final moisture content of the osmotic dehydration process. In these complementary drying experiments, the following values were used for the equilibrium moisture content and the initial moisture content: $X_i = 1.114$ (d.b.) and $X_{eq} = 0.162$ (d.b.).

In order to obtain expressions to describe shrinkage under all experimental conditions, data were turned into dimensionless values by using the following equations:

$$R^* = \frac{R_t}{R_0} \quad \text{and} \quad L^* = \frac{L_t}{L_0} \tag{52}$$

where R^* and L^* are the dimensionless values of the radius and length, respectively, at time t; R_t and L_t are the values (in meters) of the radius and length, respectively, at time t; R_0 and L_0 are the values (in meters) of the radius and length, respectively, at time t = 0.

In order to obtain the radius and thickness expressions as a function of the dimensionless average moisture content, curve fittings were performed using the Lab Fit software [38].

The expression obtained for the radius is given below:

$$R^* = 0.8712\cosh\left(0.5280\bar{X}^*\right) \tag{53}$$

Proceeding in an analogous way, an expression was obtained for the thickness of the cylinder as a function of the dimensionless average moisture content.

$$L^* = 0.7457\cosh\left(0.7865\bar{X}^*\right) \tag{54}$$

3   Results and Discussion

3.1   Osmotic Dehydration

3.1.1   Water Quantity

Optimization processes were performed for water quantity and sucrose quantity using the analytical solutions presented in Sects. 2.2.1 and 2.3.1. The results obtained for water quantity are presented in Table 1.

Table 1 Parameters obtained for water quantity in osmotic dehydration performed for a solution with concentration of 40 °Brix at temperature of 40 °C

Boundary condition	D_w (m^2 min^{-1})	h (m min^{-1})	Bi	R^2	χ^2
First kind	1.458×10^{-8}	–	–	0.9920	18.18
Third kind	1.604×10^{-8}	1.853×10^{-4}	200	0.9900	22.39

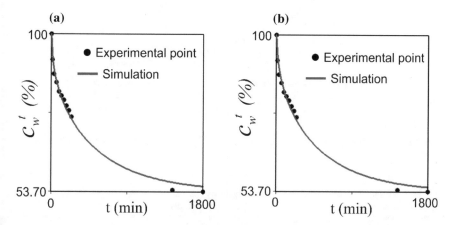

Fig. 16 Fitting obtained with the boundary condition of the: **a** first kind; **b** third kind

From the results presented in Table 1, it is possible to notice that there were no significant changes when the boundary condition of the third kind was considered. This is due to the fact that the Biot number was very large, which indicates the absence of surface resistance to water flow. In Fig. 16, it is possible to observe the fittings obtained for the two boundary conditions studied for the water quantity.

According to Fig. 16, it is possible to state that two simulations are virtually equal. Thus, Fig. 17 shows the comparison between the simulations considering the boundary conditions of the first and third kinds. In this figure a small difference is observed in the values predicted by the two simulations. This small difference of the predicted values corroborates with the values of the chi-squares presented in Table 1, which also present a small difference.

Since there are no significant differences when considering the boundary condition of the third kind, it can be concluded that the most adequate boundary condition for the water quantity is that of the first kind.

3.1.2 Sucrose Quantity

The results of the optimization processes for the sucrose quantity are shown in Table 2.

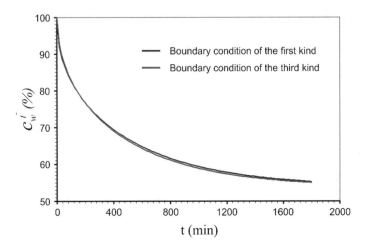

Fig. 17 Water quantity simulations considering the boundary conditions of the first and third kinds

Table 2 Parameters obtained for sucrose quantity in osmotic dehydration performed for a solution with concentration of 40 °Brix at temperature of 40 °C

Boundary condition	D_S (m^2 min^{-1})	h (m min^{-1})	Bi	R^2	χ^2
First kind	1.840×10^{-8}	–	–	0.9987	1.612
Third kind	2.039×10^{-8}	2.134×10^{-4}	181.25	0.9985	1.213

Note that the results obtained for the sucrose quantity were similar to those obtained for the water quantity, since the Biot number was very large for the boundary condition of the third kind. Therefore, the most suitable boundary condition for the description of the sucrose quantity is that of the first kind.

When the results of Tables 1 and 2 are compared, it is noted that values for water diffusivity were lower than for sucrose diffusivity. Although many authors find an inverse situation, there are several studies that point out the same behavior for the diffusivities observed in the present work: [11, 17, 39–42].

In Fig. 18, it is possible to observe the fittings obtained for the two boundary conditions studied for the sucrose quantity.

As occurred for the water quantity, in Fig. 19 a small difference is observed in the values predicted by the simulations considering the boundary conditions of the first and third kinds.

Note that the sucrose migration kinetics is equivalent for the two boundary conditions. Thus, the simplest model could be recommended to describe the process: boundary condition of the first kind.

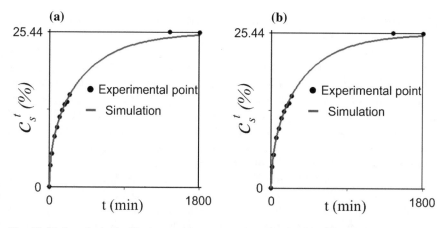

Fig. 18 Fitting obtained with the boundary condition of the: **a** first kind; **b** third kind

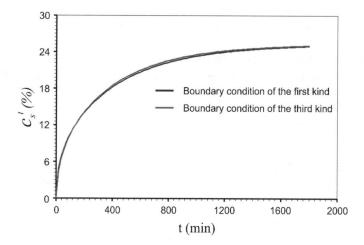

Fig. 19 Sucrose quantity simulations considering the boundary conditions of the first and third kinds

3.2 Complementary Drying

To describe the complementary drying process, the numerical solution described in Sects. 2.2.2 and 2.3.2 was used. An important phase for the beginning of the optimization process is the study of the grid and the number of steps in time. In carrying out this study, it is intended to find a grid that provides adequate results without requiring a very large computational cost. The optimization process and simulation were carried out considering a grid with 30×20 control volumes and

2000 time steps. To obtain these values, a detailed study on refinement of grid and time steps was conducted.

The results of the optimization processes for the dimensionless average moisture content are shown in Table 3. The results presented in Table 3 are related to three models: (1) considering shrinkage, variation of the effective diffusivity and boundary condition of the third kind (model 1); (2) without shrinkage, constant effective diffusivity and boundary condition of the first kind (model 2); and (3) without shrinkage, constant effective diffusivity and boundary condition of the third kind (model 3). The expression for the effective diffusivity used in model 1 was established after a detailed study of several expressions.

From the results presented in Table 3, it can be concluded that the best boundary condition to describe complementary drying was of the third kind. This conclusion was obtained by checking the best statistical indicators: models 1 and 3. Comparing the statistical indicators of the three models, it can be concluded that the best one to describe complementary drying was model 1. Figure 20 shows the fittings obtained for the three models.

The graphs shown in Fig. 20 corroborate with the results presented in Table 3. In this Figure it is possible to visualize the best fit of model 1 to the experimental data.

One of the advantages of the diffusion model over other models is the possibility of predicting the distribution of the quantity of interest inside the solid. Thus, in order to analyze the effects of the boundary condition on water distribution over time, this distribution was simulated for drying at 40 °C (Fig. 21). This simulation was performed considering the best model (model 1).

Table 3 Parameters obtained for complementary drying (at temperature of 40 °C) of osmotically dehydrated samples in solution with a concentration of 40 °Brix, at temperature of 40 °C

	Expression for diffusivity $(m^2\ min^{-1})$	Parameters and statistical indicators	Obtained values
Model 1	$bexp(a\sqrt{x})$	a	2.47 ± 0.12
		b	$(1.58 \pm 0.10) \times 10^{-9}$
		h	$(1.25 \pm 0.04) \times 10^{-5}$
		R^2	0.9998
		χ^2	7.0117×10^{-4}
Model 2	b	b	$(5.72 \pm 0.20) \times 10^{-9}$
		R^2	0.9939
		χ^2	3.7303×10^{-2}
Model 3	b	b	$(9.9 \pm 0.5) \times 10^{-9}$
		h	$(1.63 \pm 0.13) \times 10^{-5}$
		R^2	0.9992
		χ^2	4.6732×10^{-3}

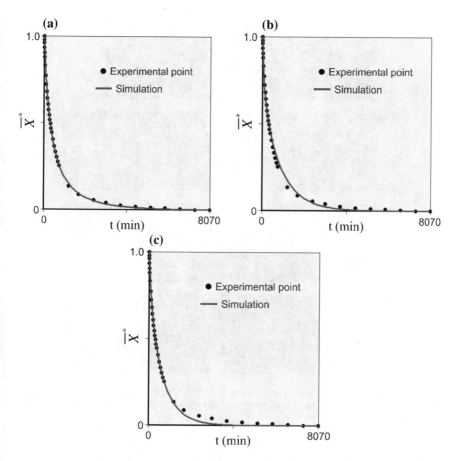

Fig. 20 Fitting obtained for: **a** model 1; **b** model 2; **c** model 3

From Fig. 21a, it is noted that after 1 h of drying the layers in contact with the medium to the north and east are not yet in equilibrium with the medium. Still from Fig. 21 it is possible to note that the equilibrium at boundary starts at about 428 min (approximately 7 h of drying).

Another way to analyze the distribution of water inside the product over time is to verify this distribution on circular surfaces (from the center of the cylinder to the top). An analysis of the distribution of water inside the product during the drying time was carried out, verifying this distribution in circular surfaces, as shown in Fig. 22.

As expected, water loss occurs from the center to the top of the product; thus, the central surface presents the highest values of moisture content at the four times

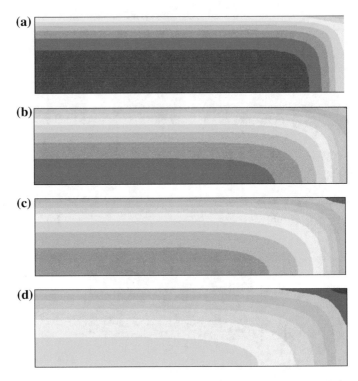

Fig. 21 Water distribution inside the cylinder (L/2 × R) for complementary drying at 40 °C (pretreated samples at 40 °Brix and 40 °C) at the times: **a** t = 60.5 min; **b** t = 181.6 min; **c** t = 250.2 min and **d** t = 427.7 min

studied. This is due to the fact that the surface in contact with the external environment loses water more quickly. As shown in Table 3, a resistance to water loss is noted both at the top of the cylinder and at the east and west layers of the cylinder. These layers begin to come close to equilibrium with the external medium at around 7 h of process.

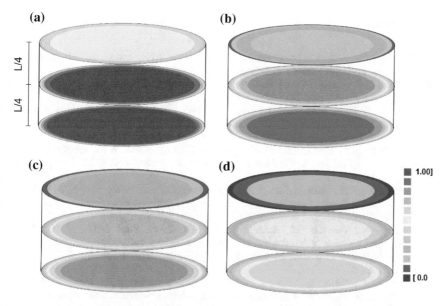

Fig. 22 Dimensionless moisture content distribution in the circular surfaces of the cylinder center, of the top and in the surface located to 1/4 of the center (drying at 40 °C of samples pretreated at 40 °Brix and 40 °C): **a** t = 60.5 min; **b** t = 181.6 min; **c** t = 250.2 min and **d** t = 427.7 min

4 Concluding Remarks

In this chapter the phenomenon of diffusion applied to processes of osmotic dehydration and complementary drying was explored. The proposed models were applied to the case of osmotic dehydration and complementary drying of banana. The proposed models proved to be appropriate to describe both the osmotic pre-treatment and the complementary drying. However, in the two cases studied, the boundary condition of the third kind proved to be important for a better description. In the case of the complementary drying, in addition to the boundary condition of the third kind, the shrinkage was also important. Finally, the models proposed in this chapter can be applied in the description of the osmotic pretreatment and the complementary drying of other cylindrical agricultural products.

Acknowledgements The authors would like to thank CNPq (Conselho Nacional de Desenvolvimento Científico e Tecnológico, Brazil) for the support given to this research and for his research grant (Processes Number 302480/2015-3 and 444053/2014-0).

References

1. Baini, R., Langrish, T.A.G.: Choosing an appropriate drying model for intermittent and continuous drying of bananas. J. Food Eng. **79**(1), 330–343 (2007)
2. Doymaz, İ.: Evaluation of mathematical models for prediction of thin-layer drying of banana slices. Int. J. Food Prop. **13**(3), 486–497 (2010)
3. Fernando, W.J.N., Low, H.C., Ahmad, A.L.: Dependence of the effective diffusion coefficient of moisture with thickness and temperature in convective drying of sliced materials. A study on slices of banana, cassava and pumpkin. J. Food Eng. **102**(4), 310–316 (2011)
4. Silva, W.P., Silva, C.M.D.P.S., SilvaJunior, A.F., Queiroz, A.J.M.: A numerical approach to determine some properties of cylindrical pieces of bananas during drying. Int. J. Food Eng. **11**(3), 335–347 (2015)
5. Falcão Filho, R.S., Gusmão, R.P., Silva, W.P., Gomes, J.P., Carvalho Filho, E.V., El-Aouar, A.A.: Osmotic dehydration of pineapple stems in hypertonic sucrose solutions. Agric. Sci. **6**(9), 916–924 (2015)
6. Panarese, V., Tylewicz, U., Santagapita, P., Rocculi, P., Rosa, M.D.: Isothermal and differential scanning calorimetries to evaluate structural and metabolic alterations of osmo-dehydrated kiwi fruit as a function of ripening stage. Innov. Food Sci. Emerg. Technol. **15**, 66–71 (2012)
7. Aires, J.E.F., Silva, W.P., Aires, K.L.C.A.F., Silva Júnior, A.F., Silva, C.M.D.P.S.: Description of osmotic dehydration of apple using two-dimensional diffusion models considering shrinkage and variations in process parameters. Drying Technology **35**(7), 815–826 (2017)
8. Germer, S.P.M., Morgano, M.A., da Silva, M.G., Silveira, N.F.A., Souza, E.C.G.: Effect of reconditioning and reuse of sucrose syrup in quality properties and retention of nutrients in osmotic dehydration of guava. Drying Technol. **34**(8), 997–1008 (2016)
9. Silva, W.P., Amaral, D.S., Duarte, M.E.M., Mata, M.E.R.M.C., Silva, C.M.D.P.S., Pinheiro, R.M.M., Pessoa, T.: Description of the osmotic dehydration and convective drying of coconut (Cocos nucifera L.) pieces: A three-dimensional approach. J. Food Eng. **115**(1), 121–131 (2013)
10. Yadav, A.K., Singh, S.V.: Osmotic dehydration of fruits and vegetables: a review. J. Food Sci. Technol. **51**(9), 1654–1673 (2012)
11. Silva, W.P., Silva, C.M.D.P.S., Lins, M.A.A., Gomes, J.P.: Osmotic dehydration of pineapple (Ananas comosus) pieces in cubical shape described by diffusion models. LWT—Food Sci. Technol. **55**(1), 1–8 (2014)
12. Silva, W.P., Aires, J.E.F., Castro, D.S., Silva, C.M.D.P.S., Gomes, J.P.: Numerical description of guava osmotic dehydration including shrinkage and variable effective mass diffusivity. LWT—Food Sci. Technol. **59**(2), 859–866 (2014)
13. Silva Júnior, A.F., Silva, W.P., Aires, J.E.F., Aires, K.L.C.A.F.: Numerical approach to describe complementary drying of banana slices osmotically dehydrated. Heat Mass Transf. (2017a). https://doi.org/10.1007/s00231-017-2120-6
14. Silva Junior, A.F., Silva, W.P., Aires, J. E.F., Aires, K.L.C.A.F., Castro, D.S.: Osmotic dehydration kinetics of banana slices considering variable diffusivities and shrinkage. Int. J. Food Prop. **20**(6), 1313–1325 (2017b)
15. Ruiz-López, I.I., Castillo-Zamudio, R.I., Salgado-Cervantes, M.A., Rodríguez-Jimenes, G.C., García-Alvarado, M.A.: Mass transfer modeling during osmotic dehydration of hexahedral pineapple slices in limited volume solutions. Food Bioprocess Technol. **3**(3), 427–433 (2010)
16. Amami, E., Fersi, A., Vorobiev, E., Kechaou, N.: Osmotic dehydration of carrot tissue enhanced by pulsed electric field, salt and centrifugal force. J. Food Eng. **83**(4), 605–613 (2007)
17. Singh, B., Panesar, P.S., Nanda, V.: Osmotic dehydration kinetics of carrot cubes in sodium chloride solution. Int. J. Food Sci. Technol. **43**(8), 1361–1370 (2008)

18. Conceição Silva, M.A.: Corrêa, J.L.G., Silva, Z.E.: Application of inverse methods in the osmotic dehydration of acerola. Int. J. Food Sci. Technol. **45**(12), 2477–2484 (2010)
19. Garcia, C.C., Mauro, M.A., Kimura, M.: Kinetics of osmotic dehydration and air drying of pumpkins (Cucurbita moschata). J. Food Eng. **82**(3), 284–291 (2007)
20. Souza Silva, K., Caetano, L.C., Garcia, C.C., Romero, J.T., Santos, A.B., Mauro, M.A.: Osmotic dehydration process for low temperature blanched pumpkin. J. Food Eng. **105**(1), 56–64 (2011)
21. Arballo, J.R., Bambicha, R.R., Campanone, L.A., Agnelli, M.E., Mascheroni, R.H.: Mass transfer kinetics and regressional-desirability optimisation during osmotic dehydration of pumpkin, kiwi and pear. Int. J. Food Sci. Technol. **47**(2), 306–314 (2012)
22. Ferrari, C.C., Arballo, J.R., Mascheroni, R.H., Hubinger, M.D.: Modelling of mass transfer and texture evaluation during osmotic dehydration of melon under vacuum. Int. J. Food Sci. Technol. **46**(2), 436–443 (2011)
23. Derossi, A., De Pilli, T., Severini, C., McCarthy, M.J.: Mass transfer during osmotic dehydration of apples. J. Food Eng. **86**(4), 519–528 (2008)
24. Zúñiga, R.N., Pedreschi, F.: Study of the pseudo-equilibrium during osmotic dehydration of apples and its effect on the estimation of water and sucrose effective diffusivity coefficients. Food Bioprocess Technol. **5**(7), 2717–2727 (2011)
25. Mercali, G.D., Tessaro, I.C., Norena, C.P.Z., Marczak, L.D.F.: Mass transfer kinetics during osmotic dehydration of bananas (Musa sapientum, shum.). Int. J. Food Sci. Technol. **45**(11), 2281–2289 (2010)
26. Mercali, G.D., Marczak, L.D.F., Tessaro, I.C., Noreña, C.P.Z.: Evaluation of water, sucrose and NaCl effective diffusivities during osmotic dehydration of banana (Musa sapientum, shum.). LWT—Food Sci. Technol. **44**(1), 82–91 (2011)
27. Luikov, A.V.: Analytical Heat Diffusion Theory, p. 685. Academic Press, Inc. Ltd., London (1968)
28. Crank, J.: The Mathematics of Diffusion, p. 414. Clarendon Press, Oxford, UK (1975)
29. Da Silva, W.P., Precker, J.W., Silva, C.M.D.P.S., Gomes, J.P.: Determination of effective diffusivity and convective mass transfer coefficient for cylindrical solids via analytical solution and inverse method: application to the drying of rough rice. J. Food Eng. **98**(3), 302–308 (2010)
30. Patankar, S.V.: Numerical Heat Transfer and Fluid Flow. Hemisphere Publishing Corporation, New York, USA (1980)
31. Tannehill, J.C., Anderson, D.A., Pletcher, R.H.: Computational Fluid Mechanics and Heat Transfer, p. 781. Taylor & Francis, USA (1997)
32. Schäfer, M.: Computational Engeineering-Introduction to Numerical Methods. Springer: Germany, 321P. (2006)
33. Maliska, C. R.: Transferência de Calor e Mecânica dos Fluidos Computacional. LTC: Rio de Janeiro, 453 p. (2013)
34. Da Silva, W.P., Precker, J.W.: Silva, C.M.D.P.S., Silva, D.D.P.S.: Determination of the application to drying of cowpea. J. Food Eng. **95**(2), 298–304 (2009)
35. Silva, W.P.; Silva, C.M.D.P.S.: "Convective" software, online, available at http://zeus.df.ufcg.edu.br/labfit/Convective.htm (2009a). Accessed 15 January 2018
36. Silva, W.P.; Silva, C.M.D.P.S.: "Prescribed" software, online, available at http://zeus.df.ufcg.edu.br/labfit/Prescribed.htm (2009b). Accessed 15 January 2018
37. Silva W. P., Silva C.M.D.P.S.: LS optimizer, version 5.1, online, available from world wide web: http://zeus.df.ufcg.edu.br/labfit/LS.htm (2017). Accessed 21 June 2017
38. Silva, W.P.; Silva, C.M.D.P.S.: LAB fit curve fitting software, V.7.2.46, online, available at: www.labfit.net (2009c). Accessed 15 January 2018
39. Amami, E., Fersi, A., Vorobiev, E., Kechaou, N.: Modelling of mass transfer during osmotic dehydration of apple tissue pre-treated by pulsed electric field. LWT **39**, 1014–1021 (2006)
40. Falade, K.O., Igbeka, J.C.: Ayanwuyi; F. A.: Kinetics of mass transfer, and colour changes during osmotic dehydration of watermelon. J. Food Eng. **80**, 979–985 (2007)

41. Khoyi, M.R.: Hesari, J: Osmotic dehydration kinetics of apricot using sucrose solution. J. Food Eng. **78**, 1355–1360 (2007)
42. Abraão, A.S., Lemos, A.M., Vilela, A., Sousa, J.M., Nunes, F.M.: Influence of osmotic dehydration process parameters on the quality of candied pumpkins. Food Bioprod. Process. **91**, 481–494 (2013)

Printed in the United States
By Bookmasters